Disease X

Kate Kelland

First published by Canbury Press 2023
This edition published 2023

Canbury Press
Kingston upon Thames, Surrey, United Kingdom
www.canburypress.com

Printed and bound in Great Britain
by CPI Group (UK) Ltd, Croydon
Cover: Daniel Benneworth-Gray

This is a work of non-fiction

MIX
Paper | Supporting responsible forestry
FSC® C013604
FSC
www.fsc.org

FSC® helps take care of forests for future generations.

ISBN
Hardback: 9781912454952
Ebook: 9781912454969

Disease X

The 100 Days Mission to End Pandemics

Kate Kelland

Canbury

CONTENTS

FOREWORD

By Tony Blair

Covid-19 was an unprecedented global crisis and should mark a turning point in global health policy and preparedness. Yet, just as we are beginning to emerge from its worst phases, we are also seeing the political will to implement the hard-won lessons we have learned melt away, and the focus on pandemic preparedness shift. In part, this is because there are several concurrent crises jostling for the attention of governments. However, health security will never be achieved if we do not build the lessons of Covid-19 into the way our governments and societies operate. To do this today, our governments need to demonstrate the same level of political will, ambition and international cooperation that leaders demonstrated in the wake of World War II when they coalesced around the objective of a sustainable peace. This must be applied to the post-pandemic order because, at its heart, health security is national security.

Disease X sets out a game-changing plan for how the world can learn from Covid-19 and be ready for the next pandemic. Global leaders should ready their nations for more frequent serious outbreaks of disease. The past few years have made it abundantly clear that we are living in an increasingly interconnected world where invisible viral threats are emerging more often, leaving all of us more vulnerable. Diseases such as monkeypox, Ebola and Middle East Respiratory Syndrome require a deep well of global willingness and leadership to prepare for and respond to.

The world must have a bold and ambitious pandemic-busting plan to ready itself for these threats.

But a plan is just a good intention – until it is delivered. In a future where we will face an increasing number of threats to human health, governments and leaders need to think about preparation not as a traditional plan-on-a-shelf that is dusted off when the threat arrives, but rather as national defence plans that are continually tested, refined and implemented across the entire system of government. This is what pandemic preparedness will increasingly resemble in the 21st Century.

While there have been severe consequences, the pandemic has also been an opportunity to transform the way we consider health. We can now make preventable disease history by building on the Covid-19 response and capitalising on the potential of next-generation vaccines and injectable therapeutics that organisations such as CEPI are developing. We have seen rapid advances in vaccine technology with new adult vaccines for malaria, Dengue and Respiratory Syncytial Virus soon to be available while countries have become adept at rapidly rolling out vaccines and the political will is growing to locate vaccine-manufacturing facilities in those areas of the world most burdened by disease. Leadership will be tested over the coming years as governments prioritise investment in these areas and facilities. They will not have immediate returns, but they will make us better prepared for tomorrow, creating more prosperous, healthier and resilient societies.

In *Disease X*, Kate Kelland lays out the dividends of decisions that are made swiftly and in concert with experts, enabling informed and decisive action. But the perils of inaction, whether in a laboratory or a cabinet room, also loom large in this arena. My Institute works deeply on these issues with governments around the world and steadfastly believes the quality of leadership matters; the Covid-19 pandemic made this abundantly clear, as Kelland so clearly elucidates. Her futuristic closing chapter outlines, through a gripping human story, what could be achieved when strong and visionary leadership is coupled with a forward-thinking agenda and thorough preparedness.

It is my hope that governments and global-health organisations maintain the momentum on tackling global-health challenges, especially in low- and middle-income countries that suffered most prominently from inequity throughout the pandemic. Such focus will result in the best return on investment from the infrastructure established for the storage, logistics and delivery mechanisms that ensured vaccines reached arms while capitalising on the catalysing research and development role that organisations such as CEPI play.

Now, CEPI has a new and ambitious goal: to compress the time taken to develop safe, effective, globally accessible vaccines against novel diseases to as little as 100 days. The 100 Days Mission is a game-changer for pandemic preparedness. *Disease X* sets out a bold and credible roadmap for how to be ready to better respond to future pandemic threats and, ultimately, to end pandemics.

Covid-19 – with its viral mutations – still endures. The long tail of the pandemic continues to suck up the ailing reserves of political will needed to address the wider challenges faced by the global community. But it is clear this will not be the last pandemic threat of our lifetimes and so we need leaders and governments to harness the opportunities and implement the lessons to be better prepared for the next one. There is no excuse to be unprepared, again.

The Right Honourable Sir Tony Blair
Prime Minister of the United Kingdom 1997–2007
Executive Chairman of the Tony Blair Institute for Global Change

INTRODUCTION

Meet Disease X

Every year in the lake-side Swiss city of Geneva, the United Nations' global health agency, the World Health Organization, convenes a committee of experts to update its long list of the most notorious nasties on our planet. These are humanity's most threatening infectious diseases. And because they both harbour pandemic potential and lack effective treatments or vaccines, they are not only the WHO's "ones to watch," but also its priorities for scientific research and development. Since its launch in 2015, the list has included some now familiar names such as Ebola, Zika and SARS. But in February 2018, the WHO added a new name – Disease X. This label, being as it is a name for something that does not yet exist, is said by the WHO to "represent the knowledge that a serious international epidemic could be caused by a pathogen currently unknown to cause human disease." In other words, Disease X represents the knowledge of what we don't know.

This era's Disease X, Covid-19, was of course unknown too, until it spilled over from the animal world – most probably originally from a bat cave somewhere in Asia – and took on the attributes it needed to infect and spread in people. It was added to the WHO's list in 2020.

But, as we have learned from studies documenting the hundreds of new human diseases that have emerged in decade after decade of the past century, Covid-19 is not the first, and won't be the last Disease X. For a start, pathogens we had once thought beaten are constantly resurging or re-emerging all over the world – from human monkeypox and malaria, to typhoid and plague. Then there are the relatively new emerging diseases – the frightening list of haemorrhagic fevers, including Ebola, Marburg, Lassa and Crimean-Congo; the deadly Nipah and Hendra bat-borne viruses; the coronaviruses that caused epidemics of Severe Acute Respiratory Syndrome (SARS) and Middle East Respiratory Syndrome earlier this century; the scary strains of bird flu and swine flu; the mosquito-borne Zika and Chikungunya viruses. Any mutant cousin of these, or any new combination of them, could be the next Disease X. And while not all new and emerging infectious diseases have pandemic potential, the next one that does could most certainly be as bad as Covid-19, if not worse.

Since the X in "Disease X" stands for everything we don't know, not knowing is something we're going to have to get used to. As people, politicians and pandemic prevention planners, we have to be honest with ourselves and admit that we don't know what's going to happen next. We can't predict what's going to emerge. Disease X is an as-yet unknown disease, about which we will continue to know very little when we first meet it. It may or may not be deadly, highly contagious and a threat to our way of life. And at this point, as well as in the months and years ahead, we also don't know when, and don't know how, it will come across the viral frontier and infect people. What we do know is that the next Disease X is already lurking, and we have to be ready for it. Just because we can't predict its features or the timing of its arrival, that doesn't mean we can't plan for it.

Meet the 100 Days Mission

It's extraordinary to think that while Covid-19 brought such turmoil and trauma, it simultaneously prompted arguably humanity's greatest ever collective scientific and public health achievement: the mobilisation of the world's vast intellectual and industrial resources to detect, understand, track, and fight a previously unknown foe. In the space of just two years between January 2020 and January 2022, we developed and built completely new defences against a completely new disease, allowing 10 billion doses of effective vaccines to be delivered to people at risk all over the world. Never before has the world vaccinated as many people in a single year against a new disease as were vaccinated against Covid-19 in 2021. And, despite the stark inequity of access we also witnessed, never before has a global vaccine rollout covered such a large percentage of the planet's population so quickly. It was the fastest vaccine development and rollout in human history.

Unlike during any previous global outbreak of a newly-emergent infectious disease, the scientific advances catalysed by the Covid-19 pandemic have also revolutionised the way the world can respond to the threat posed by future infectious disease epidemics. Thanks in part to Covid-19, we have the science, the technology and the experience to be able to develop targeted weapons against any enemy virus – known or unknown. And if we work together and act boldly, at pace, we can neutralise that virus' ability to spread and spiral from an epidemic into a pandemic. We now know we can do this. And we now know we can do it fast. Learning the lessons from Covid-19 and from history's other deadly plagues, we can build global surveillance networks that spot new threats and sound the alarm. We can prepare the emergency funding, the scientific research and the top-speed manufacturing that is needed to deploy defences against a new viral threat.

What's more, we can do so in 100 days. Stopping the next pandemic, let alone in three months, is something no single nation or organisation can do alone. But if we really want to, if we put in the sustained focus and funding needed to prepare us properly for prevention, we can ensure that Covid-19 is not only the world's worst pandemic, but also its last. We can build a pandemic-free future.

Meet CEPI, the pandemic prevention people

Our primary agent in the international pandemic-busting plan described in this book is a hitherto little-known public-private start-up organisation called CEPI – pronounced "seppy" – the Coalition for Epidemic Preparedness Innovations. I work at CEPI in London, where I am Chief Scientific Writer. I have had privileged access to it and its leaders for this inside look into how the international community is, and should be, preparing to prevent the next Disease X from becoming a pandemic.

CEPI, which is one of the architects of the 100 Days Mission, was one of the prime movers in the international response to the coronavirus pandemic. Even as much of the world had barely woken up to the threat posed by the novel coronavirus spreading from Wuhan, China, in January 2020, CEPI had already sprung into full alert, immediately initiating work on what became one of the world's largest portfolios of potential Covid-19 vaccines. Three of those are now fully licensed and being deployed to save lives around the world. Constantly on alert, CEPI's virus-watchers and pandemic prevention planners now have their eyes peeled for new emerging infectious disease threats, preparing the world to be in a dynamic state of readiness to spot and stop them before they become pandemics.

CEPI was born in part out of the experiences of the 2014-2016 Ebola epidemic in West Africa, when a tragic failure to prepare for a foreseeable infectious disease crisis ended up with it costing the lives of thousands of people and spreading to at least 10 countries. The tragedy was that while the world had begun work on a potential vaccine defence against this deadly virus, complacency about the magnitude of the threat posed by Ebola had meant it was left languishing. The world's scientific community was unable to garner enough political and financial interest to bring an epidemic-stopping tool into development. This fatal failure led to a disease – which we had seen coming and already had the know-how to defend ourselves against – being able to rip through Liberia, Guinea and Sierra Leone, spread around the world, and kill more than 11,300 people.

Even in the midst of the West Africa Ebola epidemic, and yet more so in its wake, it was clear the world needed a better way. We needed a way of coordinating and speeding the development of vaccines against threats we know are already out there, and against unknown threats we know will surely emerge. And we needed a way of addressing the euphemistically-termed "market failure" of vaccines for epidemics, where the global pharmaceutical industry is unwilling to invest in researching and developing products they might then only be able to sell in crisis-hit poor countries which can't be expected to pay high prices for them. So in 2015, Stanley Plotkin, author of the international bible of immunisation, *Vaccines*, and known as the "godfather of vaccinology," joined Jeremy Farrar, director of the London-based Wellcome Trust global health foundation, and Adel Mahmoud, the Egyptian-born American infectious disease specialist, Princeton professor and vaccine developer, in publishing a paper in the *New England Journal of Medicine*. It called for the creation of an international scientific research and development fund to develop vaccines against emerging infectious diseases. Barely two years after its conception, amid the presidents, prime ministers, CEOs and celebrities gathering at the World Economic Forum in the Swiss mountain resort of Davos in 2017, the Coalition for Epidemic Preparedness Innovations (CEPI) was born. Its mandate and vision are to develop new potential vaccines, available to all, against emerging infectious diseases, and to build a world in which pandemics no longer pose an existential threat to humanity. Central to CEPI's pandemic-preventing plan is for the world's bests virologists, immunologists and vaccinologists collectively to pre-think and complete as much of the scientific groundwork as possible before a new Disease X emerges. This will mean funding, creating and stocking a global library of vaccines against at least 100 of the more than 250 viruses that are already known to be able to infect people, and hence have what's known as "pandemic potential." The library – which I have named "GLIPP" in the final, fictional chapter of this book telling the story of how the world successfully thwarts a future Disease X pandemic – will have at least 25 sections, one each for the 25 viral families that include one or more human-adapted viruses. Having at least one, and ideally several, prototype vaccines against representative viruses for every threatening viral family will mean that when a new Disease X emerges, a prototype can be swiftly pulled off the shelf and adapted. That way, we don't lose precious and life-saving time creating totally new vaccines from scratch.

Because CEPI swiftly came to be recognised for the outsize role it played in the global fight against Covid-19, the prototype vaccine library concept, as well as the 100 Days Mission, have now been endorsed and embraced by leaders of the G7 and G20 leading industrialised nations. They, in word at least, have joined a fast-growing global consensus that we can and must invest in preparedness to be able to deploy vaccines and other biological defences against emerging diseases rapidly and collaboratively.

Back in 2017, however, the start-up CEPI's original funding – totalling $460 million – came from just a handful of forward-thinking governments – Norway, Germany, Japan and India – and two leading philanthropic funds, the Wellcome Trust and the Bill & Melinda Gates Foundation. With the European Union and British government joining as co-funders, CEPI brought together 50 or so staff, established its headquarters in Oslo, and added satellite offices in London and Washington DC. In April 2017, a few months after its launch, it recruited a softly spoken American doctor named Richard Hatchett to be its chief executive officer.

Meet Richard Hatchett

Richard is a Mississippi-born wannabe poet, turned cancer scientist, turned emergency-room doctor, turned professional pandemic worrier. In some ways, he is very much what you'd expect from a southern American gentleman. With regard to his fellow human beings, he is respectful, interested and interesting. In the rare moments when he is not working on or thinking about how to help the world prevent the next pandemic, he rides his bike in the southern English countryside, takes cultural holidays with his family, and reads books about everything from famine to philosophy to chess. His light physique and kind face, framed by soft, grey hair and striking dark eyebrows above rimless almost-not-there spectacles, draw you in. When you get to know him, one thing becomes clear almost immediately: Richard's sport is thinking. He wrestles with it, nestles in it and revels in it. And while he is distinctly humble and self-effacing, he's also quietly sure that he has substantial expertise to offer a world that wants to be able to stop infectious disease outbreaks before they spiral out of control.

A graduate of Vanderbilt University – where he majored in English, took a minor in chemistry, and had, as he puts it, "serious literary ambitions" – Richard studied medicine at Vanderbilt's Medical School before progressing to a residency in internal medicine at the New York Hospital/Weill Cornell Medical Center. There, he became interested in infectious diseases after working on a project in north-eastern Gabon investigating Ebola. Multiple outbreaks of the haemorrhagic fever had struck a tight geographical area in a tight timespan – suggesting some particular kinds of triggers might be lurking. Intrigued, Richard travelled there with a group of fellow students and spent about three and a half months studying various environmental, climatic and human factors that might have prompted these viral resurges.

So before Richard and his CEPI colleagues met their "raison d'être" – Disease X – in the form of the novel coronavirus in January 2020, he had, in his mind, already "spent a lifetime worrying about pandemics." "Who could have guessed that worrying about such rare events would turn into a career?" he said in an interview with his alma mater in 2021.

Twenty years earlier, in 2001, was the great turning point, when he volunteered on the spur of the moment to coordinate a makeshift triage field hospital for first responders at Ground Zero hours after the 9/11 attacks on New York's Twin Towers. From that moment, Richard was struck by the huge potential in harnessing the power of human ingenuity against health and other security threats. "The spirit and energy of the hundreds of volunteers who showed up was inspiring," he recalls. A subsequent invitation from then United States President George W. Bush for Richard to draw up a plan for how to galvanise medical worker might in the event of disasters set Richard on an abrupt turn from oncology into public health. Answering the President's request, he developed an original proposal for what became the Medical Reserve Corps under the direction of the U.S. Surgeon General – a force that has now ballooned into an army of more than 175,000 volunteers across the United States. Ultimately, Richard took on the leadership of global pandemic preparation after a double stint advising the White Houses of Bush and then Barack Obama on how to tackle disease epidemics and other biological threats. An important legacy from those White House days, and one that has been a guiding force for Richard throughout his career since, was a somewhat oddball group of experts that had become known as the "Wolverines." This group – named after the 1984 *Red Dawn* movie, starring Patrick Swayze and Charlie Sheen, about a gang of American teenagers who band together to defend their country from invading Soviet forces – was a self-selecting collection of dissenting doctors and public health experts, all of whom at one stage or another during the 2000s had worked together on biodefence and infectious diseases at the White House. As well as Richard and several others, the group included Rajeev Venkayya, a doctor trained in critical care who had served as special assistant for biodefence at the White House under President George W. Bush, and had written Bush's flu pandemic preparedness plan before moving to the Gates Foundation and then on to the Japanese pharmaceuticals company Takeda as head of vaccines; and Carter Mecher, an infectious disease doctor at the University of Nebraska who had also served in the White House under President Bush, and later as an adviser to President Barack Obama. As colleagues who find a spark in each other often do, the experts stayed in touch via an informal email group where, over the years, they shared their thoughts and fears about global health and disease. When a novel coronavirus emerged in Wuhan in early 2020, these biodefence intellectuals immediately revived their Wolverines "Red Dawn" email chain to share intelligence, pool their experiences, and channel late-night data-modelling frenzies into urging governments to act faster.

"The thing about pandemics is that they don't leave visible scars. They don't knock down buildings. They just fill up graveyards," Richard said in one of our many early morning conversations after I joined CEPI in 2021. Richard had recently returned from a family holiday to Croatia, Bosnia and Serbia – a culturally and historically educational trip that sharpened an insight he'd developed during years of observing and re-examining infectious disease crises. People remember wars – prompted by the daily sight of buildings pockmarked with bullet holes, lines of graves marking the premature end of hundreds of young lives, and countless memorials erected in cities and villages. But they forget about pandemics. For all the death and devastation they cause, these vast and deadly contagions of disease nevertheless fade in the collective memories of societies and generations.

In capturing retrospective attention and generating the horror that properly drives a "let's-never-let this-happen-again" assessment, the 1918 Spanish Flu has not been able to compete with World War I. And yet it claimed more than twice as many lives, killing an estimated 50 million people. How many memorials have been built to those who fell victim to this novel virus which infected their cells, flooded their lungs and turned their skin blue?

Equally, no one is likely to build memorials to the Covid-19 dead. So this book is, in part, a reminder that we should not lose sight of the Covid-19 graves, filled as they are with our friends, mothers, fathers, grandmas, uncles and siblings and so many millions of loved fellow human beings. Being reminded of those millions of lives lost should spur scientists, politicians and global public health experts to invest in and build proper defences against the many more potential existential threats lurking among emerging infections.

Meet the reality

Because, while we'd love to believe that the emergence of Covid-19 was a once in a lifetime event, that is very far from the truth. It is only a matter of time before we come face to face with the next Disease X – a new, transmissible and potentially lethal virus. Researchers identified 335 new human diseases emerging between 1960 and 2004, ranging from bird flu to the Zika virus, according to Frank M. Snowden, author of *Epidemics and Society*.[1] And since the 1980s, the number of infectious disease outbreaks in people around the world has tripled. It takes little more than a flutter of a genetic mutation to turn a pesky local disease outbreak into a pestilent global plague.

The threat of another major pandemic – one that could easily be more deadly and catastrophic than Covid-19 – is higher than ever.

Yet the horrific toll of at least 6.5 million lives lost,[2] billions of infections, and 28 trillion dollars wasted shows that governments around the world were not prepared for this decade's Disease X. Among policymakers and publics ill-equipped to fathom exponential viral spread, infectious diseases had come to be seen as old-fashioned threats – problems of bygone eras when sanitation was poor and scientific knowledge was paltry. Historically, governments have tended to hide behind a view that infectious disease crises are one-off events – and when this or that year's vicious virus, or this or that region's pernicious pathogen, goes away, the event is over and we all go back to normal.

What this does not grasp is that infectious diseases are emerging even more frequently than they did in the past. We've often heard the Covid pandemic described as a once in a century event. But such descriptions – rendering disease pandemics freak, rare and "out of our control"-type events – give false comfort. The far less comforting reality is that emerging pathogens are a truly modern threat – as new and dangerous as cyber-attacks, extreme weather events and other systemic threats to our way of life. And far from being a once in a lifetime or century event, Covid-19 is the seventh global infectious disease crisis of the 21st Century. Starting with SARS in 2002-2003, we've seen H5N1 bird flu emerge in 2004-2005, H1N1 "swine flu" in 2009 – which became a pandemic – then MERS in 2012, Ebola in 2014-2016 and Zika in 2015-2016. Each of these could be argued to have been a "Disease X" of its time. What's clear is that these are not freak events, but events that are happening continually, and at an increasing cadence.

I'm writing this book while the emotional scars left by the Covid-19 pandemic are still raw, and while graves are still being filled with victims of this era's Disease X. That's because the time when we are mourning, reflecting and rebuilding is precisely the time to understand how to prepare ourselves, our governments and our societies to view and deal with infectious disease epidemics as *a routine part of modern life*. This book lays bare the reality of a world where that drum beat of emerging infectious diseases echoes with increasing regularity as each month and each year passes. It challenges us to test our collective state of readiness and suggests we identify and hone certain key traits to help us get mission ready. To build a pandemic-preventing ecosystem that will allow us to swiftly face down future pandemic threats, we need to be prepared to be scared, to act quickly on those fears, to be bold and take risks; we need to know that we will get some things right, and some things wrong, and that in doing so we will move forward even as we suffer setbacks; we need to get our heads around how we can plan ahead for something we cannot predict. Above all we need to know that preventing pandemics is possible, as long as we are properly prepared. As Larry Brilliant, an American epidemiologist who helped the world eradicate smallpox, is often quoted as saying: "Outbreaks are inevitable, but pandemics are optional." If we don't embed the knowledge we have now about how to get ahead of infectious disease epidemics before they spiral out of control, we will be caught unawares and unprepared once more.

1: PREPARE to be Scared

The start of 2020 was Richard Hatchett's Cassandra moment. From the first few hours of the New Year, with the emergence of a cluster of cases of a mysterious pneumonia in Wuhan, China, he sensed impending doom. He'd seen emerging viruses before. He'd seen coronaviruses before. He'd seen pandemics before. At this time, he was uncomfortably certain that this newly-identified pathogen was something to be taken deadly seriously.

Like the mythological Cassandra, the prophesying princess who foretold the fall of the great, ancient city of Troy, Richard began to cry out warnings. To anyone he felt he could talk to straight, he did. In the first few days and weeks of January 2020, as others around the world kicked off their new-planned resolutions and wished friends and loved ones a Happy New Year, Richard sent texts, emails and WhatsApps to colleagues, friends and family members with a very different tone. The "novel coronavirus," as it was then known, is "really scary," he told them, and would "be everywhere in the next few months." He warned that the spread of the virus and the disease it caused was probably already out of control and would "cause tremendous social disruption." The outbreak that had started and spread from a distant province in China already looked to him, he said, like the "worst infectious disease crisis in my life – and that includes two pandemics and HIV."

These warnings elicited mixed reactions. Richard's wife Susan, as well as his three children – two of them teenage and one still at primary school – were a little unnerved for sure. They were ready to take on his advice for precautionary steps. Heeding his warnings, the family began buying in small supplies of masks, medicines and non-perishables. Richard's colleagues were more than just a little freaked out. What worried them, however, was not so much the threat of the novel infectious disease itself, but more Richard's state of extreme alarm. "He was running around with his head on fire," was the description offered by Melanie Saville, Richard's friend and colleague at CEPI, where he is – and was then – her boss. "My personal perception was that it was really over the top – really quite extreme," added Dick Wilder, CEPI's United States-based General Counsel. Wilder recalls how he watched, at times aghast, from across the Atlantic as Richard, in urgent conversations both within his professional expert circle and beyond, applied war-time analogies to the looming disease threat and warned that the new contagion could cause tens of millions of deaths.

Yet Richard had learned a lot over the years about what fear can do – starting with an intense experience at Ground Zero in the days following the terror attacks on the United States on 11th September 2001. When two passenger planes slammed into the twin towers of the World Trade Centre in downtown New York, Richard was working as an emergency room doctor at the Memorial Sloan Kettering Cancer Centre, about five miles away at the other end of Manhattan Island. Strange as it may seem, his emergency room was quiet that day. "We spent the entire day preparing to receive patients who ultimately never arrived," Richard recalls. "People either made it home, or died."

With few or no patients making it to him, and with the prospect of a few days off shift at the ER anyway, Richard decided to go to where he might find some people he could help. On the day following the attacks – Wednesday 12th September 2001 – he made his way towards the smouldering wreckage of the Twin Towers (later named Ground Zero) to volunteer his assistance. After hitching a ride with several other would-be helpers on a truck that was driving down the West Side Highway, he got out a few blocks north of Ground Zero and checked in at Stuyvesant High School. The school had been evacuated after the World Trade Center attacks and was becoming an ad hoc meeting point for medical volunteers wanting to see how they could help. There, as the smoke and dust from the destroyed Twin Towers still filled the air, catching in peoples' throats and stinging their eyes, Richard soon found himself among hundreds from the medical community and beyond who wanted to offer whatever assistance they could. These were people with purpose – people with expertise, energy and empathy – all desperate to put it to use. Richard says he particularly remembers finding a strong sense of value and motivation among the volunteers who had shown up. Everyone was energised and ready to act, he says.

But what he also found was a frustrating level of chaos. No organisation. The urgent call had gone out for people to come and help, but little thought had been given to how their expertise and empathy could be harnessed and put to work. So, when the volunteers heeded the call and showed up, there was no system or structure there to receive or deploy them. It was as though no-one had thought anything like this could ever happen.

Because he had some ER experience, Richard found himself drawn into taking over an informal leadership role. When the by now exhausted volunteer who had initially been trying to coordinate the growing numbers of willing but largely directionless helping hands decided to go home and get some rest, Richard stepped in. With fear as a catalyst, Richard witnessed a torrent of human goodwill and other resources pouring in, and he set about guiding it. Within hours, the little triage area had expanded into a four-story field hospital serving the hundreds of first responders who had been deployed to the rubble of the Twin Towers site. It soon also began to attract millions of dollars' worth of donated medical equipment and emergency drugs that neither Richard nor any of his co-volunteers had asked for. It became, within a day or so, the critical hub for providing medical support to the fire, medical and search and rescue workers toiling around the clock in the devastation of Ground Zero.

"The profound insight I had coming out of those extraordinary few days at Ground Zero is what you can accomplish in terms of coordinated human effort when you're galvanised by an external threat," Richard recalls. "It was a personally transformative moment for me. Very powerful. When you have a galvanised community, even in the midst of something terrible, people can accomplish amazing things. People can do things even beyond their sense of their own capacity. They can do anything. It's not something you can write into a policy plan, but it *is* something you can count on."

Although they are a very different kind of external threat, emerging diseases do, by their obscure nature, induce fear. At their starting point – which in our terms is when they first emerge and are identified in human populations – they are characterised by invisibility, uncertainty, unfamiliarity and uncontrollability. Examples of modern-day emerging diseases – such as the human immunodeficiency virus (HIV) that causes AIDS, and the deadly Ebola haemorrhagic fever – have each come with a sizeable dose of fear attached. And historical examples – such as the "Black Death" pneumonic plague in the 14th Century and the 1918 "Spanish Flu" – are now recognised as having caused the deadliest pandemics in recorded human history. Outbreaks or epidemics of emerging diseases add yet another frightening and often incomprehensible element: the potential for rapid and even exponential spread among populations often described by scientists as "naive." In other words, people (or animals) whose immune systems have never before encountered a pathogen and hence have no knowledge of how to fight it. While "R nought," also known as the "basic reproduction number" of a disease, has become a household phrase in recent years, the idea of exponential spread of a disease was little better understood among politicians and the general public in 2020 than were concepts of "lockdown," "furlough," or "messenger RNA." In epidemiology – the science of the spread of disease – R nought is crucial to being able to tell how bad things could get, and how quickly. At the start of any novel disease outbreak, one crucial uncertainty is the emerging disease's R value, or the average number of people an infected person will, in turn, infect. What *is* known, though, is that any R value above 1.0 means exponential growth – in other words, it means something is increasing at a faster and faster rate..

Added to this unknown but explosive potential is the fact that many infectious disease epidemics are caused by viruses. The word virus derives from the Latin for "slimy liquid poison." It was first used in the medical community in medieval times to describe the liquid, pussy discharge from an infected wound or ulcer. Around 75 per cent of emerging infectious diseases in humans originate from animals – and virology experts estimate there could be more than a million undiscovered viruses currently lurking in wildlife, any or all of which have the potential to jump species and attain the traits and capabilities needed to infect and spread among people. As well as being invisible, viruses – unfortunately – are also everywhere. Planet Earth is home to more viruses than to all other life forms combined. Scientists estimate there are around 100 million types, each of them a disarmingly tiny, but also potentially deadly, piece of genetic material wrapped up in a cloak of proteins.

Despite the enormity of the threat some of them can pose to humans, viruses are actually minuscule. In February 2021, Kit Yates, a mathematician at Britain's University of Bath, captured headlines when he calculated that all of the Covid-causing virus circulating in the world at that time – when there had been almost 107 million confirmed cases of Covid-19 globally and the disease had killed more than 2.34 million people – could fit easily inside a single cola can. Using global rates of new infections with Covid-19, coupled with estimations of viral load in each infected person, and with the diameter of the spiky, ball-shaped SARS-CoV-2 virus – at an average of about 100 nanometres, or 100 billionths of a meter – Yates worked out there were around 200 million billion SARS-CoV-2 virus particles in the world at any one time. While that's a big number, its volume in virus terms is tiny, at barely more than 160 millilitres.

Viruses are both the smallest of all microbes and, at the same time, the ultimate parasite. Viruses only exist to make more viruses. But because they cannot survive or replicate alone, they instead infect a host's cells and hijack components of it to generate energy to make copies of themselves. Those hosts can be any living thing – any type of organism from the largest mammal to the smallest plant to the teeniest of micro-organisms, including bacteria.

Yet despite their tiny nature and their inability to do anything on their own, viruses can be extremely dangerous. In the process of infecting and hijacking the cells of a host, they are entirely selfish. What they leave behind – the remnants of the cells they have purloined for their own replication – is often severely damaged, and sometimes dead.

One of the most fearsome features of viruses is that – in part because of their invisibility and unfamiliarity – they can be profoundly disruptive of human patterns of sociability. They don't always bring out the best in people, that's for sure. And they can trigger alarmingly swift and alarmingly negative changes in behaviour. We only need to look back at the emergence of HIV and AIDS to find how the fear of this novel killer quickly morphed into a fear of the people carrying it, often coupled with disgust and prejudice about their social and sexual preferences. The Ebola epidemic of 2014 to 2016 in West Africa also saw fear of the virus quickly turn into fear of the closely-associated, newly-erected healthcare facilities and treatment centres designed to help those infected. This fear of going to a clinic or hospital arguably exacerbated the spread of the deadly Ebola disease within communities.

So viruses can be both biologically and morally pernicious. They can turn people against each other and foster misplaced fear. That an inert piece of genetic material can have that kind of power should be something that frightens us all.

In early 2020, several small but important clues to the emerging disease causing severe illness in Wuhan began to become apparent through a series of publications and announcements between 7th January and 11th January.

On 8th January, the World Health Organization (WHO) said the cluster of more than 50 pneumonia cases in the Chinese city might be due to a newly emerging member of the coronavirus family. This was the same family that had previously spawned viruses that caused the deadly SARS outbreak in 2003 and the yet more lethal outbreak of Middle East Respiratory Syndrome, or MERS, in 2012.

The following day, Chinese state television gave a little more detail, citing the findings of a preliminary assessment of test results: "As of 7th January 2020, the laboratory detected a new type of coronavirus," China Central Television reported, adding: "The new coronavirus that caused this epidemic situation is different from previously discovered human coronaviruses." Then, on 11th January, the genetic sequence for the novel coronavirus was published on an open-source website called virological.org.[3] Later that same day, China's Centers for Disease Control (CDC) also sent the genetic sequence to the WHO, which successfully urged Beijing to make it public.

For disease detectives and vaccinologists, this fascinating data dump of genetic code – a detailed sequence made up of mesmerising blocks of four-letter combinations – was a central clue. It allowed them to begin work on deciphering whether or not the pathogen was similar or different to previously-encountered coronaviruses, and in which ways; how it was getting in to human cells; how it was replicating there; and what might be done to stop it.

For epidemic predictors and pandemic worriers, the release of the genetic sequence turned some crucial unknowns into knowns, giving them tiny pieces of knowledge amid the fog of an unfolding event: this was indeed a respiratory disease; it was caused by a coronavirus; and it had never before been seen in humans. This knowledge sharpened their fears that something like SARS – which killed almost one in 10 of those it infected – or MERS – which had an even more dreadful death rate of 35 to 40 per cent – was coming to get us again. And it also left many unknowns still unknown.

The clues to deciphering these further still uncertain features were to come as a different kind of data began to build up: epidemiological data – information on the numbers of people infected, on when and where they picked up the virus, and on whether they became a little sick, severely sick or died. Information coming out of China at this time was sporadic.

By 13th January, China was reporting 80 confirmed cases of infection with the novel coronavirus. A week later, the number had exploded by more than 2,000 per cent to 1,860 confirmed cases by 20th January. Another week on, the number had jumped again to 12,426 cases – a weekly increase of more than 600 per cent.

These numbers began to answer questions raised by our previous experiences with the coronaviruses that caused SARS and MERS. While the MERS virus has a horribly high mortality rate of up to 40 per cent, it has – so far – not shown the capability to spread easily between people. This is a factor that has – again, so far – kept a cap on the MERS virus' pandemic potential. Its earlier-born coronavirus cousin SARS, on the other hand, was a little more transmissible, but also a little less deadly. It was eventually contained after killing around 800 people in more than two dozen countries. Would this next novel coronavirus have all, or just some, of these capabilities?

After the week between around 13th January and 20th January, when the number of cases of infection with the new coronavirus began to escalate rapidly day after day, so too did the numbers of people who were dying from the new disease. The two deaths reported in the week of 20th January rose by more than 2,500 per cent to 53 deaths in the week ending 27th January. The following week, the death toll had jumped by more than 500 per cent to 306. This was a deadly exponential outbreak in action. And this was also the moment at which attuned epidemic-watchers began to be sure that what had emerged and was rapidly spreading in China was a pathogen with pandemic potential.

While the numbers still looked relatively small, the pace at which they were growing was frightening. This outbreak already belonged in the category of a "clearly highly communicable (disease that was) clearly going to infect billions of people," Richard says. And when you start multiplying any kind of death rate – even if it's a half or one per cent – by a billion, it swiftly becomes clear you're talking about an awful lot of dead people. "My estimate from mid-January was that we were in an event that, unmitigated, could potentially kill 50 million people," he says. On his Red Dawn email chain, his fellow Wolverine pandemic-watching experts were also afraid – in part because of the numbers, and in part because they knew most political and public health experts were likely to fail to see the enormity of what was coming. "Maybe disease outbreaks need a warning like the one on your car mirror," wrote one of the leading Wolverines, Carter Mecher, a former intensive care doctor who had also worked at with Richard at the White House in the early 2000s. "Things are much larger than they appear." So while, in other places around the world, hundreds of millions of people largely ignored what was going on in Wuhan and got on with shopping the January sales or planning their year ahead, Richard began stocking up on facemasks, basic foodstuffs and over-the-counter medicines for his kitchen and bathroom cupboards, and advised others to do the same. "We have ordered 50 face masks a person for our family plus a small supply of N95s and are doing some prudent provisioning of non-perishables," he wrote in a WhatsApp message to his friend Arturo in January 2020. "Just wanted to convey my best advice."

Richard was not alone in busying himself and those closest to him with precautionary actions. In the city of Wuhan, the epicentre of the outbreak of the novel coronavirus, construction engineers were racing to build and complete a makeshift emergency hospital to treat patients infected with the new disease. China was aiming to construct the facility – built over two floors and including 30 intensive care units and several isolation wards – in a matter of days so that it could begin admitting up to 1,000 patients suffering with severe illness from Covid-19.[4] The project brought back memories of 2003, when – in response to the emergence of SARS – the Xiaotangshan Hospital was built in the Chinese capital, Beijing, in seven days, reportedly breaking a world record for the fastest ever construction of a hospital. As news of the Wuhan coronavirus outbreak spread in January 2020, international media again jumped on the hospital story – fascinated by the speed of the construction and wooed by the record-breaking nature of Chinese effort. A headline on the BBC News Online website asked: "How can China build a hospital so quickly?" and quoted experts citing Chinese authoritarianism and the regime's "top-down mobilisation approach" as factors that helped it get resources into such projects at such speed.

Like the world's media, Richard's interest was piqued by the enormously accelerated efforts Chinese authorities were piling into building hospitals and other emergency facilities. But, unlike the headline writers and politicians, he wasn't interested in *how* they could embark on and complete such ambitious healthcare infrastructure projects so quickly, or *how* they could mobilise so many people and so much money to plough into them. He wanted to know *why*.

This different approach was, in part, a result of lessons Richard had learned from what he describes as his "intense engagement" with the writings of Central and Eastern European dissidents in the 1980s. As a curious and internationally-minded American growing up in the southern United States, Richard had enthusiastically consumed the work of Milan Kundera, the exiled Czech writer and author of *The Unbearable Lightness of Being* whose main character, Tomas, was a surgeon.[5]

This was "my favourite novel in those years," Richard says, in part because it had taught him to observe and think about nations in new ways. Most importantly, he says, it taught him to understand that a nation's people and its politics are two entirely different things and should be viewed as such. To confound the two, Richard learned, was to read a situation totally wrong, missing what could be critically useful insights.

"I saw that kind of confounding of politics and life in the way the world was reading China's response to SARS-CoV-2," he said. "There was a lot of political and ideological discounting of China's actions. There was a tendency for people to bring their biases, to say these things were happening 'because this is an authoritarian government', 'because they're communists' or 'because they don't have any respect for human rights' and so on." In essence, many politicians and commentators in the West were turning it into a political narrative about the Chinese government and its ideology, rather than observing and recognising it as a public health response to something that was really frightening. While China was scared, and responding to that fear with decisions it hoped would mitigate the threat, much of the rest of the world was slipping into a narrative that justified inaction.

"My reading by that point was that we were in a pandemic," he recalls. "And part of what enabled me to read the situation correctly was that I read what the Chinese were doing as the response of a community that had been through this kind of thing before – with SARS – to its first encounter with another totally new virus. I didn't buy into any of the political overlay. The political overlay spoke to what the Chinese could accomplish by building a hospital in five days, yes. But the fact that they were building a hospital in five days had nothing to do with Chinese communism, it had everything to do with what they were fighting."

This recognition of a nation's fear, and a refusal to judge it or explain it away with political or ideological differences, added to Richard's sense of alarm, and further encouraged him to continue in his Cassandra role even in the face of concern and dismay among some of his colleagues. In CEPI's offices in London – on the first floor of the Wellcome Trust's Gibbs Building on the busy Euston Road in central London – Melanie Saville remembers Richard showing her online webcam video footage of the empty streets in Wuhan. It was "really eerie and scary," she recalls, and in showing the footage Richard was, in some ways, trying to make sure she saw and felt that fear. "At about the same time, Richard took me aside and said: you know the virus is probably already in the UK, right? There's no stopping it."

Richard's fear was not personal, but global – a dread that events would unfold in ways that humanity was unable to deal with, and in ways that would pit region against region, country against country, citizen against citizen. "What was behind my sense of alarm was recognising that by mid-January we had probably already lost control of an event that was going to become a gigantic global pandemic. That was really scary. I understood that it probably didn't present a super high risk for me personally, but globally it was shaping up into an unbelievable catastrophe."

You may question why fear should be part of a 100 Days Mission aimed at neutralising viral threats and mitigating epidemic risks. What can preparing to be scared do to help us on our way? The answer is that we should not fear the fear. Being frightened is the very thing that *sets* us on our way. It is the trigger for us to react. Without it, we would do nothing.

Playing the prophet of doom, or at least sounding the alarm, sat easily with Richard. He has often described himself as having "made a career out of worrying about pandemics." Unlike most people who, especially in the pre-Covid era – and totally understandably too – have not developed a capacity to think about pandemics and the threats they present in a strategic and constructive way, Richard had by then already spent a couple of decades doing exactly that.

His experience at Ground Zero in 2001 – and most importantly his realisation of the gains that could be made by turning an ad hoc call for volunteers into a structured emergency response – had prompted him to write a memo to some influential public health specialists. They in turn, sent it to the White House, which at that time was occupied by President George W. Bush. Richard's note proposed that some kind of national army of medical volunteers should be established and put on stand-by so that it could step up whenever the next public health crisis hit. A few weeks later, Richard got a call from then Vice President Dick Cheney's office to come to Washington and put his plan into action, creating the U.S. Medical Reserve Corps that today has a headcount of more than 175,000 volunteers.

Now within the U.S. administration's circle of threat-response thinkers, and with a place at the table of President Bush's Homeland Security Council, Richard went on to analyse a range of possible threats to U.S. security, from nuclear radiation to bio-terrorism attacks using the smallpox virus. From this position, Richard was also able to witness the effect on President Bush of fear prompted by existential threats. The 9/11 terror attacks were swiftly followed by a series of anthrax attacks on media outlets and politicians. Outside of the United States, but no less threatening, were the first human outbreaks of H5N1 bird flu, with a terrifying case fatality rate of 60 per cent, and the emergence of SARS in 2003. Then, in 2004, America suffered an acute shortage of flu vaccines due to a contamination problem and delivery suspension at its major supplier in Britain. And in 2005 came hurricane Katrina. "I think the combination... really frightened him," Richard says of the then President's mindset. By September 2005 – after a summer break during which Bush had reportedly read a book by John Barry called *The Great Influenza : The Story of the Deadliest Pandemic* – the President's fears were sufficiently heightened for him to call on his emergency response advisers – Richard and Carter Mecher among them – to draw up a pandemic preparedness plan.

These White House years – repeated again when Barack Obama was elected President and called on Richard and others to advise him on the 2009 H1N1 swine flu pandemic – meant that by the time China decided to put Wuhan into lockdown, Richard had imagined his way, over and over again, into what explosive global health crises and fast-moving deadly pandemics can look like. He'd theorised and planned for the first sporadic cases becoming exponential spread. He'd envisioned disrupted societies and locked-down economies. He'd figured out how and when interventions like work-from-home orders, school closures, quarantine and self-isolation would be most likely to be effective. He'd also foreseen the kinds of death tolls that could soon become daily headline news. In all these imaginings, he had plotted and modelled and mapped and projected what could happen if a highly transmissible novel respiratory virus with killer potential, even a low percentage case-fatality rate, could not be contained before becoming a pandemic. "Knowing how small the window was for the world to act to prevent it, this (novel coronavirus spreading in China) scared the bejesus out of me," he recalls.

During those same early hours and days of January 2020, at its headquarters in Geneva, the WHO was frantically trying to get information out of Chinese authorities about the reported cluster of atypical pneumonia cases in Wuhan. On 1st January, the agency activated its Incident Management Support Team (IMST), a part of its emergency response framework designed to ensure public health emergency response activities are coordinated across the three levels – global, regional and national.

By 5th January, the WHO had enough information to be able to share some details about the cases in a note distributed via its Event Information System, which is accessible to all of its 194 Member States. The note advised national governments to take precautions, but did not raise high alarm.

The snowballing of events over the next few weeks – including the reporting by Chinese media on 11th January of the first death from the novel coronavirus; news of the first recorded lab-confirmed case of infection with the virus outside of the People's Republic of China – in Thailand on 13th January; and a steep rise in the number of cases and deaths to almost 2,000 and more than 50 respectively worldwide by 20th January – increased pressure on the WHO to convene a meeting of the International Health Regulations Emergency Committee on 22nd and 23rd January. Such meetings are called when a disease outbreak is on the horizon that looks as if it has the potential to "go global". They offer a chance for the WHO to sound an early warning – a declaration of a Public Health Emergency of International Concern, or "PHEIC" (pronounced by some as "fake") – signalling that the world's health security is under threat.

The WHO has only sounded the PHEIC alarm seven times since the system was put in place as part of updated International Health Regulations in 2005. Before the Covid-19 pandemic and the 2022 monkeypox outbreak, these were when: H1N1 (swine flu) emerged in Mexico and spread to the United States, ultimately becoming a pandemic; polio resurged in Afghanistan, Pakistan and Nigeria in 2014; an Ebola outbreak exploded in 2014 and spread throughout Guinea, Sierra Leone and Liberia; an epidemic of Zika virus started in Brazil in 2016 and spread internationally; and another deadly Ebola outbreak erupted in 2018 and spread in the Democratic Republic of the Congo.

Yet in the third week of January 2020 – despite having evidence of international spread of the virus reported in the Republic of Korea, Japan, Thailand and Singapore, of its ability to kill, and of the possibility of human-to-human transmission even in asymptomatic cases – the Emergency Committee was split. Unable to agree on the global seriousness of the threat to human health, its advice was that the outbreak in China and beyond did not at that moment constitute a PHEIC. This tied the hands of the WHO's Director General, Tedros Adhanom Ghebreyesus, who could do little more than take the advice and refrain from declaring an international emergency.

Infectious disease experts around the world were baffled by this, and said so publicly. Peter Piot, the Belgian scientist who helped discover and first identify the Ebola virus, a professor of global health, and at that time director of the London School of Hygiene & Tropical Medicine, issued a statement saying that "regardless of the decision not to declare this a Public Health Emergency of International Concern" the world needed to act together urgently to have a chance of stopping the outbreak. "There cannot be any complacency as to the need for global action," he said. While other public health experts were openly angry at the WHO's decision, Richard was, at the very least, frustrated. On the evening of 23rd January, just a few hours after the WHO's announcement that, for now, there was no global health emergency, he wrote to his friend and confidant Jeremy Farrar: "All I can say re PHEIC is: what in God's name is it for then?"

To be sure, the balancing act such global agencies need to perform when seeking to put the world on alert while not triggering undue panic is delicate in the extreme. It's a matter of sounding the alarm as loud as you possibly can without creating chaos and confusion, and without losing credibility. It's difficult to get right. For the WHO, it was a week later, after a second teleconference meeting of its Emergency Committee lasting five hours, that the announcement that many felt was overdue eventually came. On 30th January 2020, the WHO declared, in what Farrar, for one, described as a "belated acknowledgment of what many of us already feared", that the novel coronavirus outbreak was indeed a Public Health Emergency of International Concern. It wasn't until almost three months after that, on 11th March, when there were more than 118,000 cases globally and almost 4,300 deaths, that the World Health Organization began using the word "pandemic" to describe the global outbreak of Covid-19.[6]

Starting this book with the idea that being scared is an essential ingredient for being successful in beating pandemics may well raise eyebrows. Is fear really such a good thing? Can it not paralyse us? Make us freeze – like proverbial rabbits in headlights? Can feeling fear not very easily slip into spreading it? Is "playing the Cassandra" not just a polite way to describe fear-mongering or crying wolf? Can a continuous banging of the drum of fear be dangerous for societies and their people – making them anxious, on edge, insular, cowering?

The answer to these is, at times, yes. Raising fears does carry the risk of having these negative effects. But it also brings critical benefits.

Scientists who have studied fear describe it as a primitive and powerful emotion which, in its most primary function, serves to alert us to the presence of danger or the threat of harm. Fear comes from the brain. When we're frightened, a part of the brain called the hypothalamus reacts by releasing a series of chemicals that kick-off the so-called "fight or flight" response – a way in which all our animal senses are put on high alert for us to react quickly.

The story of the fear that triggered some experts' and even some governments' initial reaction to the emergence of the SARS-CoV-2 virus tells us something about the importance of this primitive feeling in preparing for pandemics. If we are not afraid, if we don't sense danger, or if we choose to ignore, dismiss, explain away or even ridicule the fear we see others experiencing, we miss a vital trigger that spurs exactly the sorts of actions we need to take to mitigate a threat.

Does that mean, then, that to be properly prepared, we need more people to be more afraid? Again, the answer to this is – on balance – yes. If the arrival of Covid-19 has taught us anything, it's that the risk of pandemics is a permanent feature of modern life. It's here to stay. And it's one we need to be suitably afraid of. Like cybersecurity threats or the ever-present risk of international conflict. Governments, business and citizens should think about how the world can protect itself against new and re-emerging pathogens in the same ways it seeks to protect itself against invasions or cyber threats. Take the malicious Stuxnet computer worm cyberweapon. It was originally aimed at Iran, where it destroyed a number of centrifuges in a critical uranium enrichment facility, and was then modified by various groups to attack scores of other facilities including gas lines and water and power plants. While the threat from Stuxnet, as from some biological viruses, has lessened over time, partly due to our response to it and efforts to block it – our collective thinking about it and cyber threats has not. Thankfully. "We don't think about computer threats as, 'Oh, Stuxnet, well it's gone now – we have the patch, so we don't need to worry about cyber security anymore'," Richard says. And just like those other modern-life risks of cyberweapons and conflict, pandemic threats are not going to stop being a constituent feature of our lives just because we have only recently seen the risk turn to reality. The Covid-19 pandemic's occurrence in 2020 does not mean we're not at risk of more.

This means that for us be able to prepare properly, and ultimately to neutralise that risk, we need to accept and process the threat so that it both scares us, and then informs and guides our behaviour. This is not to suggest everyone everywhere should live in constant fear, paralysed by anxiety about potential deadly threats. Not at all. It's a question of internalising the threat and making mitigation part of modern living. Just as we would never set up new phone or computer systems without spending serious time and money installing firewalls and anti-virus scanners to protect us from cyberattacks, so we should not assume that just because we're just emerging from the worst infectious disease crisis in a century, we can rest assured we don't need to protect ourselves against the next one. We do. It is vital, if we're going to be able to act quickly and effectively in the face of new emerging viral threats – to succeed in our mission to build effective human defences in just 100 days – to start by fully recognising and accepting that they are part of the reality of modern life. And that they are frightening. "It's a threat and a risk that everybody needs to internalise," says Richard. "And it's one that those who aspire to political leadership especially need to internalise."

And so it was that a few weeks into February 2020 – on Valentine's Day in fact – Richard was out for dinner and, in his words, "playing a slightly tipsy Cassandra." This was no romantic date night with his wife Susan. This was something else. The venue was the Watteau Salon at the Bayerische Hof Hotel. The host was Erna Solberg, Norway's then Prime Minister and Head of State. The context was the Munich Security Conference – an annual February gathering of the world's defence and security elite in the Bavarian capital – designed as a forum for discussion and, ideally, diffusion of the greatest global threats of the moment.

Around the table were the great and the good of global public health – Seth Berkley, chief executive of the GAVI Vaccine Alliance, Chris Elias of the Bill & Melinda Gates Foundation, Caroline Schmutte of the Wellcome Trust global health fund. Beside each place setting was a small note – a bit like a flyer that would advertise an upcoming film or poetry reading or concert – detailing the discussion that was to be had. It read: "From Ebola to Coronavirus: how can world leaders boost global health security." As is customary at such events, the pre-dinner scene setting was done via a series of introductory speeches by each of those attending – a sort of "tour de table" where each person trots out a rehearsed mini-speech that, as Richard puts it, "extols the verities of international cosmopolitan multilateralism, and makes them look clever."

Richard was in no mood for such niceties. By this time, he was even more frustrated at the lack of global alarm about the novel coronavirus. It was already three or more weeks since he had been at the World Economic Forum in Davos and had there signed three separate deals to invest in developing new vaccines against this new and dangerous pathogen. Disinhibited by a little more red wine than he'd usually drink at such events, and fuelled by his ever more certain view that the world was – right now – facing potentially the greatest global health security crisis in recent history, he let loose. "I was like: 'Hey! People! You know, talking about the verities of cosmopolitan multilateralism – that's REALLY not what we need to be doing right now! This is a pandemic! It's coming! It's coming fast and it's going to sweep over us like tsunami! We all need to understand that it's coming now and it's coming fast and it's not going to stop!'," Richard recalled. "OK, I wasn't quite shouting or banging the table or anything, but I was definitely not sticking to any pre-rehearsed speech. Instead, I was sitting there in front of a Head of State and saying: 'listen, we all need to be very, very afraid!'."

2: PREPARE to Move Fast

When Stéphane Bancel, the French businessman, biological engineer and chief executive of the now well-known pharmaceutical firm Moderna, emailed Richard on the evening of 20th January 2020, it was to ask him to place a lightning-fast almost-blind, one-million-dollar bet. Bancel, who has since been made a knight of the Légion d'Honneur in France for his achievements, was at that time – a cold and nervous start to the year – feeling a little less than fresh. He'd just flown in on the seven-hour overnight red-eye Boston to Zurich flight to travel up into the Swiss Alps to join other chief executives, world leaders and decision-makers at the World Economic Forum in Davos. He was, however, nonetheless eager to get going on a scientific development he thought could become a world first. And he needed money, fast, to do so. "We were anxious to get moving," Bancel recalled afterwards. "We had known about the virus since between Christmas and New Year, but without the genetic information we could do nothing. But then came a moment that we could start moving."

Richard, who was in Davos to be part of an expert panel discussion about the unfolding outbreak in Wuhan, China, had already clocked and calculated the deadly potential threat hidden in the small cluster of pneumonia cases being reported there. When he got Bancel's email, it took him just seven minutes to decide what to do. It was a go: "Short answer is I think so," Richard wrote back in an email from his iPhone. Within 48 hours, his words became cash. A contract to provide Bancel's biotech start-up Moderna with just shy of $1 million was signed on the night of 22nd January, allowing the U.S. company to immediately start test production of an experimental vaccine against a virus that was still so new it had not yet been named. "He just needed money to manufacture it," Richard recalls. "So I said 'sure!'."

"Starting early is the name of the game, because you can never make up for lost time," says CEPI's U.S. director and strategic advisor Nicole Lurie, reflecting on those first super-fast decisions as the novel coronavirus was only just emerging. "We had a deep appreciation of the need not just to move science fast, but to move money fast, and the need to be deadly serious about it. You can't wait for somebody else to send up a balloon or pull a trigger, because by the time they get comfortable enough with decision-making to pull the trigger, it's way too late."

Making quick, decisive moves in the early stages of a rapidly spreading disease outbreak is a prerequisite for getting ahead of a potential pandemic. But doing this is made all the more difficult by what is often likened to the "the fog of war" – that critically uncertain period where we know a major threat is upon us, but we know barely anything about its nature, potential or consequences. In some senses, disease outbreaks are not unlike wars. They're scary, uncertain and often seem senseless. When the world starts to see what can happen when a new viral foe emerges, jumps species, mutates and starts to spread among people, we recognise it as a fight for lives, for freedoms, for our way of life. But the "fog" that swirls around those first few hours and days can drastically delay progress. The dilemma faced by scientists and public health experts tasked with battle-planning – including those at the World Health Organization seen in Chapter 1 – is that moving swiftly in a fog is very difficult. And no matter which way and at what speed you do manage to move through the fog, you're liable later to come in for criticism from those you're trying to protect. The WHO was condemned for moving too hastily to declare the 2009 H1N1 "swine flu" outbreak a pandemic – although much of this criticism was levelled at the United Nations agency with hindsight, when the disease had proved far milder than many had feared and its death rate had proved thankfully lower than predicted. Yet when Ebola erupted in West Africa in 2014 and went on to cause the largest epidemic of that disease in history, the WHO was criticised for being dangerously slow to react. An independent panel of experts convened by the Harvard Global Health Institute and the London School of Hygiene and Tropical Medicine which analysed the WHO's response to the 2014 Ebola outbreak even went as far as recommending it should be stripped of its role in declaring international emergencies arising from disease outbreaks. One expert on that panel, Professor Ashish Jha, director of the Harvard Global Health Institute, said the delay in sounding the alarm was the WHO's "most egregious failure," adding: "The cost of the delay was enormous."[7]

By the time Covid-19 hit the world and was recognised as a global pandemic threat, the new director of the WHO's health emergencies programme, the ruddy-faced straight-talking Irish infectious disease epidemiologist Mike Ryan, was acutely aware of the need to advise national governments to move quickly: "Be fast. Have no regrets. The virus will always get you if you don't move quickly," he said. "Speed trumps perfection."

Ryan was, in part, criticising those national governments which had hit out at the WHO for supposedly holding back on details as to how, how far and how fast the novel coronavirus was spreading. Ryan was pointing out that at the same time as accusing the WHO, many governments were taking little or no firm action themselves. In Britain, where the first cases of Covid-19 were reported on 31st January 2020, it took another two months, until 23rd March, for a full national lockdown to be imposed. While Britain's then Prime Minister Boris Johnson was undoubtedly ramping up the rhetoric in early March – warning in a speech on 12th March that "many more families are going to lose loved ones before their time" and that the spread of Covid-19 was "the worst public health crisis for a generation" – his actions, or lack of them, spoke louder than his words. Far from moving swiftly to close schools and non-essential shops, his most dramatic act was to advise against international school trips, and to suggest that the sick, the frail and the elderly should avoid going on cruises.[8] Jeremy Farrar, head of the Wellcome Trust and a former scientific adviser to Johnson's government, said in an interview with *The Times* when he stepped down from his advisory role that he hadn't fully anticipated "just how slow, insular and unprepared governments would be to understand what was happening and respond effectively."

In the United States things moved even more slowly. With even more deadly consequences. An analysis conducted by STAT in conjunction with researchers at Britain's Oxford University, published in June 2020, found that a direct comparison of the earliest stages of the pandemic – when all governments and policymakers were dealing with similar levels of fog and uncertainty – showed fatal slowness in the United States. The analysis calculated that in the four months following the first 15 confirmed cases of Covid-19 in the United States, 117,858 Americans died of the disease. After the same period in Germany, by comparison, only 8,863 people died – in large part because Germany acted significantly faster in appreciating the seriousness of the threat. Modelling a scale-up of Germany's population of just under 84 million to a hypothetical 331 million to match the United States population, the analysis found that if American leaders had acted as swiftly as the Germans, 70 per cent of those 117, 858 deaths – or around 82,000 of them – could have been prevented.

Some of the world's more dynamic governments included New Zealand, South Korea, Germany, Singapore, and, the source of the outbreak, China itself. Experts estimated that the initial Wuhan lockdown, imposed on 23rd January 2020, prevented between 0.5 and 3 million infections and between 18,000 and 70,000 deaths in the city in this very early stage of the novel coronavirus outbreak. In New Zealand, where the first Covid-19 case was reported on 28th February 2020, it was just two weeks before severe travel restrictions were put in place, and three weeks before the government announced a four-level national alert system. Three and a half weeks after New Zealand's first case on domestic soil – on 25th March 2020 – when the case count had grown to 205 but there were still no Covid-19 deaths, Prime Minister Jacinda Ahern implemented one of the strictest lockdowns in the world, only allowing people to go out for absolute essentials like food and medical care. Professor Michael Baker, a doctor and expert in public health at the University of Otago who devised New Zealand's response and advised its government on what to do when, reported being struck by a WHO report on the success of Wuhan's lockdown. It persuaded him that it was right to act fast, he said, and, despite the fog, to "throw everything at the pandemic at the start."

Inside CEPI's HQ in London, Richard's recognition of the need for speed – something that allowed him to make, among others, that seven-minute decision on providing almost a million dollars to seed-fund Moderna's experimental Covid-19 vaccine – came in part from a deep academic analysis he and colleagues had conducted back in 2006 and 2007. The study dealt with the (inappropriately-named) Spanish Flu – the novel influenza virus that swept across the world in 1918 in the midst of and in the wake of World War I. Working with Marc Lipsitch, then a professor in the Departments of Epidemiology and of Immunology and Infectious Diseases at the Harvard School of Public Health, and with his newly-found White House friend, Carter Mecher – on secondment from his position at the Department of Veterans Affairs – Richard took a detailed look back at the deadly 1918 pandemic.

In particular, the researchers' interest was in looking at the effect of 19 types of so-called "non-pharmaceutical interventions" or NPIs – things like school, office and factory closures, social distancing, and voluntary quarantining of infected households – in American cities during the autumn phase of the Spanish Flu pandemic. Going back over old newspaper reports, historical documents and academic papers, the doctors found that different cities and local authorities had taken social and public health policy decisions at different times. And that had made a difference. Because one thing was especially critical in determining whether NPIs had a great, or not-so-great, impact: timing. The social distancing measures – in particular things like closing schools, churches, dance halls, saloons and theatres – were far more effective in curbing the spread of the killer flu virus in cities that introduced them early and rapidly. Peak death rates in cities that started intervening swiftly were around 50 per cent lower than those in cities where authorities acted more slowly. In Philadelphia, for example, where the first cases of Spanish Flu were reported on 17th September 1918, school closures, a ban on public gatherings and other social distancing measures were not introduced until two and a half weeks later, on 3rd October. The cumulative death rate in the city reached 719 per 100,000 people. In St Louis, meanwhile, where the first reported case was on 5th October 1918, social distancing measures were introduced just two days later. There, the cumulative death rate of 347 per 100,000 people was less than half of Philadelphia's. Not only that, but fast-acting local officials were able to flatten the epidemic curves in their cities if they moved both to introduce NPIs early, and to keep those measures in place for longer.

While the paper may have seemed to many to be little more than an academic exercise, even an unnecessary re-examination of a 100-year-old flu pandemic that had surely been examined over and over already, Richard knew instantly in January 2020 that its findings had life-saving potential. "Carter, I am seriously thinking this may be the event for which the work we did on NPIs may become critical," he wrote in an email to his friend and co-researcher on 22nd January 2020, the day before Wuhan went into lockdown. Dr Anthony Fauci, the in-equal-measure maligned and celebrated Chief Medical Adviser to former U.S. President Donald Trump who became a household name during the Covid-19 pandemic, said at the time of the study's publication: "A primary lesson of the 1918 influenza pandemic is that it is critical to intervene early."

Yet in 2009, when swine flu emerged in South and then North America and began spreading swiftly around the world in a WHO-classified pandemic, Richard's and his fellow pandemic-watchers' repeated reminders about the need for speed had still not filtered through. In advice to the then United States President Barack Obama, Carter Mecher, Richard and the response team called for immediate action, asking the president to put into action a pandemic strategy they had been instrumental in drawing up during the previous – Bush – administration. The strategy proposed closing schools, encouraging social distancing and other non-pharmaceutical interventions in what

Richard describes as a "no-regret set of decisions while we could gather more information about the outbreak." But experts at the U.S. Centers for Disease Control (CDC), the national agency charged with guarding America's public health, advised Obama to hold fire and wait for more data – in other words, to see whether more confirmed cases, and possibly deaths – would show up. Obama went with the CDC, and delayed.

As luck would have it, the 2009 H1N1 swine flu pandemic turned out to be remarkably mild, with a death rate of not more than 0.01 per cent. According to CDC data, some 60 million Americans were infected with the H1N1 flu over the course of a year, but the disease killed only around 12,500 of them. Which meant, of course, that hindsight framed the delayed response by the Obama administration and other slow-moving governments around the world as clever, considered and correct. Not so, argues Richard. In another scenario such delays could be catastrophic. The world just got lucky that swine flu was relatively mild.

But luck is not a strategy, and no such luck was to soften the deadly blow of sluggish government reactions to the arrival of the Covid-19 pandemic in 2020. Richard watched in horror and frustration as political leaders once again played "wait and see." "It was the same response (as in 2009): let's wait, let's collect data," he said. "But while in 2009 it didn't matter because it was a mild virus, in 2020 it was a tragedy."

Just like their federal counterparts, American city authorities had not learned the speed lesson either. In New York, it was on 1st March 2020 that a 39-year-old woman tested positive for the novel coronavirus after flying home to the United States from Iran a week earlier. By this time, many of the world's populous cities and countries had evidence a-plenty that the novel coronavirus was extremely contagious, and that any local cluster of cases could quickly balloon into a major outbreak. But New York's authorities chose rather to brag and bluster about the city having some of the best hospitals in the world and being able to handle this thing so much better than other places. On 2nd March, the city's then mayor, Bill de Blasio, encouraged New Yorkers in a tweet to "go on with your lives + get out on the town despite Coronavirus."[9] It was another two weeks before de Blasio ordered schools to close, on 15th March, and another week beyond that before New York State authorities issued a stay-at-home order. "Everything was slow," Stephen Levin, a Brooklyn Democrat serving on the city's council, told *The New York Times* afterwards. "You have to adapt really quickly, and nothing we were doing was adapting quickly."

While few knew about it at the time, CEPI was one of the Covid-19 pandemic's fastest movers. The seven-minute decision to fund Moderna's first batch of experimental Covid-19 vaccine was abruptly followed by two more lightning-fast investments, so that by 23rd January 2020, when just 581 cases of the infection caused by the novel coronavirus SARS-CoV-2 had been confirmed worldwide and when the disease didn't even have a name, the coalition had already kick-started three investments into potential defences against it.[10] Those projects, launched with an American biotech company called Inovio, with Moderna in collaboration with the U.S. government's National Institute of Allergy and Infectious Diseases, and with a team of vaccinologists at the University of Queensland in Brisbane, Australia, were among the very first in the world to begin work on developing Covid-19 vaccines. It's notable that all three of these super-fast start-up projects were collaborations with vaccine research scientists whom CEPI had picked out way before Covid came along, way before the novel coronavirus had made its mutating journey from a bat cave in China to a wet market in Wuhan and on into its first handful of human hosts. Richard's relationship with Moderna's Stéphane Bancel, for example, dated back to his days at the U.S. government's Biomedical Advanced Research and Development Authority, or BARDA, when as Chief Medical Officer and Deputy Director he had signed an agreement with the then tiny biotech company in 2016 to use its mRNA technology to start developing a vaccine against Zika.

With Inovio too, CEPI had some history. In 2018, CEPI gave Inovio up to $56 million over five years to further its work on potential vaccines against the Middle East Respiratory Syndrome (caused by another coronavirus) and Lassa – a viral disease first identified in 1969 that causes potentially deadly cases of haemorrhagic fever across West Africa. CEPI's tie-up with the University of Queensland began in 2019 with a $10.6 million award for a scientific team at the university to develop a so-called rapid response vaccine platform – a technology designed as a "plug and play" option for creating new vaccines. "We knew we needed to act quickly just in case, so we built on work we had been doing before," explains Melanie Saville, CEPI's Executive Director of Vaccine Research and Development. "We went to those partners and said 'please start working on this now. Build a vaccine candidate, get it into the clinic (clinical testing). You need to hit the ground running. Every day you wait is a day lost'."

Think also about the Oxford-AstraZeneca coronavirus vaccine – one of the world's fastest-to-be-developed, most effective and most accessible Covid-19 vaccines – whose co-developers Professor Sarah Gilbert and Dr Catherine Green have zoomed to global vaccinology rock star status (and, in Gilbert's case, also earned a damehood under the British honours system). CEPI's decision in March 2020 to give catalytic funding to help pay for the making and testing of starting materials for what eventually became the Oxford-AstraZeneca Covid-19 vaccine was not the first step in that relationship either. A year and a half earlier, in September 2018, CEPI had struck a deal to provide up to $19 million to the Oxford group to develop vaccines against Lassa, Nipah, and MERS, a type of coronavirus. This meant that, crucially, by the time the next novel coronavirus, SARS-CoV-2, had come across the viral frontier in 2020, the Oxford scientists' ChAdOx1 platform with MERS had already shown promise in so-called Phase 1, or early stage, human trials.

Ask any pacey achiever – be they sportsperson, entrepreneur, technician or academic – how did you do that so fast? How did you get there so quickly? And they are likely to answer with something like this: I knew exactly what I needed to do, I was extremely well prepared, I'd practised over and over and over again, and I got off to a great start. Being prepared and able to move fast is made all the more achievable by having as much as possible done in advance, and by knowing in deep detail the intricacies of what needs to be done, as well as why, when and how.

Just like a pit stop in Formula One motor racing. This complex, multi-person operation to replace four wheels and refuel a highly-tuned racing car in the midst of a high-stakes competition used to take well over a minute. Video footage of a pitstop team for the American Formula One driver Bill Holland at the Indianapolis 500 race in 1950 shows a handful of mechanics in baggy overalls and baseball caps laboriously hammering away at one wheel nut after another to remove the worn-tyre wheels and replace them with fresh ones. Another mechanic polishes the windscreen, another shuffles over to bring the driver a drink in what looks like a regular teacup, while two more man-handle the fuel hose and pump to refill the car's petrol tank. It all takes an excruciating, nail-biting 67 seconds! Not long at all, in the grand scheme of things, yet still so much slower than it could be. Seventy or so years later, Formula One pitstops at Grand Prix races around the world feature a team of 20 or more technicians poised like coiled springs as the car is driven at top-speed into a centimetre-accurate stopping place. Each mechanic completes one critical task and the car is refitted and ready within less than two seconds. Even for those with little interest in the sport itself, this 97 per cent reduction in pitstop time cannot fail to make an impression.

This relentless focus on pace is just as critical – indeed more so – in managing disease outbreaks. Because, as Richard's 2006/7 academic study of NPIs showed, and as the responses of Wuhan, South Korea and New Zealand made real, the quicker an outbreak of a contagious disease is contained, the lower the threat it poses to lives and livelihoods. Every day counts.

Much like the two-second pitstop, the 100 Days Mission that forms the core of how the world can respond effectively to a future outbreak of infectious disease and contain it before it becomes a pandemic is built on the principle that speed can and must be embedded into every stage of the response. Starting with fast alerting systems to sound the alarm and trigger quick decision making, governments and global health bodies then need to move swiftly to implement containment measures like school closures and work-from-home orders to the keep the outbreak in check. That then gives scientists and pharmaceutical developers the time they need to design and create new medicines, and most importantly, new vaccines to defend people against the virus.

Without doubt, vaccines are our most potent tools against pandemic threats posed by known and unknown viruses. They have been an answer – sometimes only in part, and sometimes, sadly, too late – to calls on the scientific community over the last decade to respond to epidemics of H1N1 swine flu, Ebola and Zika. They have been critical in controlling Covid-19 – saving tens of millions of lives and preventing hundreds of millions of cases of severe illness that would otherwise have completely overwhelmed and crushed health systems across the world. And they will be critical to ensuring future outbreaks of infectious disease are stripped of their pandemic potential before it becomes real. The faster an effective vaccine is developed and deployed to the most vulnerable populations, the faster an emerging infectious disease pandemic threat can be contained and controlled.

Vaccines can take many years – even decades – to develop. Take Mosquirix, for example, the world's first malaria vaccine, which took 25 years to move "from lab to jab," as they say in the vaccine development world. Mosquirix's chief developer, a bearded, softly spoken, deep-voiced Egyptian-American research scientist called Joe Cohen, announced the success of his vaccine at an international conference on malaria on 18th October 2011. It was a fabulous day, Cohen told me in an interview at the time. But it came after "many ups and downs, and moments over the years when we thought 'Can we do it? Should we continue? Or is it really just too tough?'." Results of the vaccine's testing, which at the time was Africa's largest-ever clinical trial (Mosquirix was tested in around 16,000 children over several years across seven countries on the continent), showed that it halved the risk of African children in the trial getting malaria. Cohen said he had begun work on the vaccine as a naive young molecular biologist in 1987 – on April Fool's Day in fact, 1st April. His boss at a drug company then called Smith, Kline & French – which subsequently became GlaxoSmithKline, or GSK – asked him if he would step in to lead the firm's malaria vaccine research programme. It was just after an early-stage experimental vaccine had failed in tests, managing to protect only one of several vaccinated volunteers in the trial. Taking on the job to lead the programme from there, Cohen said, entailed "quite a bit of soul searching." "I felt I was taking on an enormous scientific challenge," he recalled.

Getting on for a quarter of a century later, Cohen said the 2011 trial results made him feel fabulous. "This is a dream of any scientist – to see your life's work actually translated into a medicine... that can have this great impact on peoples' lives." He admitted, however, that he had "never dreamed it would take this long."

Thankfully, recent history of vaccine development shows that science is making accelerated progress. Before the SARS-CoV-2 virus emerged in late 2019 and triggered a rush to develop Covid-19 vaccines at record-breaking speed, the previous vaccine development record was for a live attenuated mumps vaccine. The work, led by the legendary American microbiologist and vaccinologist Maurice Hilleman, took just shy of five years – or almost 1,800 days – from the point at which he isolated the virus in March 1963 to his vaccine being licensed for use in January 1968. The first stage of this development – involving multiple experiments and tests of dose sizes and immune responses, first in cells in petri dishes and then in animal experiments, before a potential vaccine can even be tested for the first time in a human clinical trial – took two years in itself.

During the outbreak of Severe Acute Respiratory Syndrome, or SARS – now often referred to as "SARS One" – in 2003, U.S. scientists took around 20 months to progress from having the genetic sequence to the first human trials. By then, the SARS outbreak was already petering out naturally, with the virus not spreading easily from person to person. By comparison, it took 326 days from the identification of the SARS-CoV-2 virus to a Covid-19 vaccine being brought into emergency use. That represented a quantum leap in pace. And the speed record for the pre-clinical trial stage of vaccine development was 66 days, set by Moderna.

An analysis published in 2022 by Amanda Glassman, Charles Kenny, and George Yang at the Center for Global Development in Washington DC[II] found that for previous vaccines for which data on timescales are available, the average period between microbe isolation and vaccine development was 48 years. The same analysis found that the average time it took from developing a vaccine to achieving 40 per cent coverage with that vaccine was 42 years – with a record of 14 years for a vaccine against rotavirus, a highly contagious virus that can cause life-threatening diarrhoea. For Covid-19, the period between vaccine development and 40 percent coverage was just 11 months. The volumes have been astounding too. More than 7 billion doses of Covid vaccines alone were manufactured and administered in 11 months. That's around 50 per cent more again than the roughly 5 billion doses of all types of vaccine the world would normally – in other words, before Covid – produce in a whole year.

It's clear that we're now in an historic moment for vaccinology – one that means that future pandemics can be optional, as Larry Brilliant put it, rather than inevitable. Some of the dramatic difference in speed of vaccine development is down to an explosion in basic scientific understanding, including in areas like genomics and structural biology, that has fuelled new advances in vaccine development. Since 1977, when the British scientist Frederick Sanger sequenced the first full genome – that of a virus called phiX174 – gene sequencing speeds have accelerated at an eye-watering pace. The Human Genome Project, launched in 1990, took 13 years to identify, map and sequence all of the genes in the human genome. By 2020, viral genomes could be sequenced in one or two days.

Some of the difference, too, is down to the urgency of the threat and being suitably alarmed to start working on things fast, as we saw in Chapter 1. And crucially, it's about being primed: knowing who's working on what, why, and where it might have cross-over relevance to the current crisis so help get us ahead. What's clear is that the pace the world set in developing Covid-19 vaccines didn't come from nowhere. The scientific innovation didn't come out of the blue. Just like other record-breaking human achievements – be they Usain Bolt's 9.58 second 100m sprint, Dick Rutan's and Jeana Yeager's first non-stop flight around globe, or the Apollo 11 first moon landing – the Covid-19 pandemic's game-changing scientific advances were built on years of hard graft: learning, training, experimenting and trialling. Remembering the January 2020 email exchange with Richard during Davos, Moderna's Stéphane Bancel reminded an audience at a Global Pandemic Preparedness Summit at London's Science Museum in March 2022: "We had been working on this technology for 10 years." The Canadian scientist, Oxford University Professor and UK government advisor Sir John Bell, speaking at the same summit, noted that the Covid-19 pandemic had spawned "a plethora of really exciting scientific developments." But none of the successes were things that had started from scratch. All the things that worked, he said – be they successful vaccine technologies like the mRNA constructs or viral-vector platform that formed the basis of the Oxford-AstraZeneca shot, steroid drugs like dexamethasone, or even the lateral flow and PCR tests – were all things that scientists and developers had, to a greater or lesser extent, already got their heads around before the novel coronavirus emerged. "All the work that went on *before* is absolutely crucial," Sir John said.

It's striking that among the most successful programmes set up to create and develop Covid-19 vaccines, many had code-names based on pace. "Operation Warp Speed" in the United States was modelled on America's mobilisation efforts during World War II and aimed to deliver 20 million doses of at that time not yet developed vaccine doses by December 2020. The German firm BioNTech, which teamed up with Pfizer to develop an mRNA vaccine against Covid-19, dubbed its coronavirus pandemic endeavour "Project Light Speed." Embedding the need for speed into the plans seemed to pay off. The Pfizer-BioNTech vaccine later claimed a slew of records: the first Covid-19 vaccine to be authorised by a globally-recognised regulator for emergency use; the first also to be cleared for regular, non-emergency use; the first fully tested and stringently authorised Covid-19 shot in the world to go into an arm – that of the then 90-year-old British grandmother Margaret Keenan, who got her jab at her local hospital in Coventry at 6.31am on 8th December 2020, less than a year after the first case of the new pandemic disease was diagnosed.

When Dr William Kaelin, the Nobel Prize-winning scientist, gave a speech during the Covid-19 pandemic for medical students graduating at Johns Hopkins in the United States, he referred to the coronavirus crisis as an "accelerant – accelerating many developments that would have taken decades to happen, if they happened at all." The crucial thing now is for the scientific and public health community not to lose that record-breaking momentum. When the next Disease X emerges, the world needs to move faster still than it did in response to SARS-CoV-2 if the disease's threat is going to be neutralised before it becomes a pandemic. Moving on from the likes of Warp Speed and Light Speed, I like to call this "Pandemic Speed" – a next level pace of scientific and societal progress that will take the world ahead of infectious disease outbreaks so that we can safely defend ourselves and our way of life against them. The 100 Days Mission, proposed and outlined by CEPI in early 2021, subsequently backed by leaders of the G7 group of wealthy countries and supported by G20 governments, is going to need this next-level Pandemic Speed. It is a plan to squeeze the time it took for Covid-19 vaccines to be developed – already a history-making 326 days – down to just 100 days. That's a little over three months. That's fast.

In an interview in the summer of 2021 about the 100-day vaccine idea, Eric Lander, a geneticist, microbiologist, mathematician and former science advisor to President Joe Biden, put it like this: "It makes you gulp for a second, but it's totally feasible."

Totally feasible. And totally life-saving. Imagine we had developed a Covid-19 vaccine within 100 days of the genetic sequence being made available to the world. The first people in the world to be vaccinated – the likes of Margaret Keenan and of the then 81-year-old Bill Shakespeare, the first man to get a stringently licensed Covid jab – would then have been given their protective injections in April 2020 rather than in December 2020. That would have been when there were just 2.3 million confirmed Covid-19 cases in the world, rather than eight months later when there were more than 68 million cases of infection. Faster vaccine development and deployment would not only have saved many of the millions of lives lost to Covid-19, it would have also prevented economic damage running into trillions of dollars. Think too, of the way the various mutant variants of Covid-19 developed and brought massive deadly second, third and fourth waves of the pandemic to countries before vaccines had become widely available. If transmission of the SARS-CoV-2 virus had been less widespread and less rampant – kept in check by a vaccine rollout that had started more than half a year earlier and had hence needed to contain a far smaller epidemic – would the likes of Delta, Omicron, BA.4 and BA.5 ever have had the chance to develop? The longer a virus is able to circulate in a species, the more time and opportunity it has to develop small changes – mutations – in its genetic code. A speedier start, earlier use of non-pharmaceutical interventions, and swifter delivery of effective vaccines might all very well have starved Covid-19 of at least some of the many chances it had to evolve and spit out new variants – variants that caused wave after wave of sickening and deadly epidemics and in turn forced vaccine developers to have to rethink and adapt their approach to a constantly moving target.

That speedier start, the ability to get off the blocks at Pandemic Speed in the future, means having a dedicated global surveillance network that scans high-risk pathogens and can swiftly spot the next Disease X. A global early warning system. And this is all the more crucial in an era when new outbreaks are likely to crop up more frequently thanks, in large part, to human activity driving climate change, deforestation, urbanisation, and rapid, repeated, constant international travel. These human activities are in turn driving increasing rates of zoonosis, or "spillover" – when viruses circulating in one or more animal species evolve and jump into a person and then mutate into a transmissible form that humans can easily pass to one another. Many parts of the surveillance network are already in action scanning both people and animals for emerging new pathogens. In a programme called PREDICT, funded by the United States' Agency for International Development (USAID) an army of almost 7,000 virus hunters in 35 potential spillover hotspot countries collected more than 140,000 biological samples – mostly extracted from the blood, saliva, faeces and urine of bats, primates and rodents.

In the 10 years between 2009 and 2019, their detective work led to the identification of 1,200 viruses with the potential to cause human disease and pandemics, including – guess what – more than 160 new coronaviruses, and a previously unknown strain of Ebola. A follow-up project launched by President Joe Biden's administration in 2021 called DEEP VZN, or Discovery & Exploration of Emerging Pathogens – Viral Zoonoses, is now working in targeted countries in Africa, Asia, and Latin America to explore in more detail the types of viruses and animal hosts that are most likely to provide a spillover threat, and to assess the level of risk.

Then there's the Global Virome Project – a scientific endeavour to map and genetically sequence the hundreds of thousands of viruses circulating in animals and birds that come from viral families already known to be able to infect humans – families like the coronaviruses, which gave us SARS, MERS and now Covid-19, and the filovirus family, which spawned the Ebola virus and the equally frightening Marburg virus, both of which cause viral haemorrhagic fevers in people. Space science is also getting in on the virus-hunting action. NASA satellites that have for years been monitoring daily fluctuations in the earth's environment – including things like rainfall, vegetation coverage and changes in climate – have already shown they can pick out areas where there is a looming threat of a potentially deadly infectious disease outbreak such as malaria or Rift Valley fever, both of which are carried by mosquitoes and transmitted to people in their bites. An early warning system using NASA data successfully predicted an outbreak of Rift Valley fever in 2006 and 2007 in north-east Africa six weeks before the first people started to get infected. The crucial thing for getting ahead of the next Disease X is to build, link up and maintain a global early warning system that will "leave no stone unturned," says Gabrielle Breugelmans, CEPI's leading expert in epidemiology, the study of the spread of disease. "To avoid the next pandemic, we've got to move fast, and we must always be on the lookout."

It's compelling to think by how many days, or weeks, the world could have got ahead of the novel SARS-CoV-2 coronavirus when it first emerged and began to spread in people. And then to imagine too that Pandemic Speed had been injected into the entire global response from the very start of the outbreak in Wuhan. A study by academics at Britain's Southampton University WorldPop research programme took a look at what the Covid-19 outbreak might have become if authorities in China had introduced non-pharmaceutical interventions like quarantining, social distancing and lockdown measures a week, two weeks, or even three weeks earlier than they did in January 2020.[12] They found that making a move just seven days earlier could have reduced the number of cases by 66 per cent. And, according to the Southampton research team led by Professor Andy Tatem and Dr Shengjie Lai, taking another week or another two weeks off that time to act could have led to a reduction in cases of 86 per cent and 95 per cent respectively. Getting ahead by just three weeks, and using nothing more technical than stay-at-home orders, quarantining, school and non-essential business closures, would have meant that by the end of February 2020, there would have been fewer than 6,000 cases of Covid-19 in China, rather than the estimated 114, 325. Imagine if China had done everything it did in mid-January 2020 in mid-December 2019 instead. And that Covid-19 vaccine development had started in December 2019 rather than late January 2020. And that the scientists and drug developers had then delivered their vaccines in only 100 days, rather than in more than 300 days. If, accepting that they have all been demonstrated as possible, you add those things together, there is a real prospect that the world could have controlled SARS-CoV-2 when it was still an outbreak and hence avoided the Covid-19 pandemic it became. A real prospect that we could have reacted so fast, moved at such speed and worked with enough urgency to have contained and neutralised the SARS-CoV-2 virus' pandemic potential. It's a sobering thought.

3: PREPARE to Take Risks

The signs had been extremely promising. A dedicated team of expert vaccinologists at the University of Queensland in Australia had been working around the clock on a ground-breaking technology. Their hope was to produce a cutting-edge type of vaccine that could, potentially, help protect their country's entire population – and more beyond – from the deadly virus that was sweeping the world in a pandemic. The work was excitingly intricate. Professors Paul Young, Trent Munro, Keith Chappell and their team had invented a new type of mechanism that was able to stabilise the viral protein that forms the basis of a vaccine – in this case, the notorious 'spike protein' of the novel coronavirus – and lock it into shape. This fixing allows the immune system more easily to recognise and identify the virus, and hence to mount an immune response to it. Vaccines developed using this technique – known to aficionados as a "molecular clamp" – are likely to be more effective and able to be manufactured very rapidly in large volumes. Clearly, the potential upsides of rapid, large-scale, highly effective vaccine production in a global pandemic were game-changing.

It's not surprising then that the University of Queensland project caught CEPI's eye. Since the opening days of January, CEPI's Executive Melanie Saville and her research and development team had been relentlessly scanning the global vaccinology scene for new and different approaches that might produce swift, effective defences against the new viral enemy. Inspired by positive early results, CEPI struck a deal – announced on the same day as the Moderna and Inovio agreements – to give Young's team $4.5 million to help them advance to the next stage of clinical testing as quickly as possible. Also optimistic that this approach might be a breakthrough, the Australian government, led by then Prime Minister Scott Morrison, put in a one billion Australian dollar advance purchase order for 51 million doses. Boosted by CEPI's support and racing against the unfolding coronavirus pandemic, Young's team dedicated every waking hour to the project. They created their first Covid-19 vaccine candidate in the lab within just three weeks. It was a great start – and an exemplar of Pandemic Speed in action.

But 11 months later, the exhilaration among Young's team members turned to devastation as they realised it was time to pull the plug on their once-so-promising Covid-19 vaccine development project. The molecular clamp chosen for its enhanced stabilising properties was made using two fragments of a protein found in HIV – the human immunodeficiency virus that causes AIDS. The team of scientists had thought, with good reason, that the risk of these tiny fragments causing an HIV immune response was very low. Ultimately, however, they were wrong: volunteers in the early-stage clinical trial of the Covid-19 vaccine candidate began showing up as positive in HIV tests. And while there was no suggestion at all that the vaccine was actually giving them HIV, the risk of hundreds, if not hundreds of thousands, of false positive HIV tests in vaccine recipients around the world was just too high for the project to be allowed to proceed.

Dick Wilder, CEPI's top lawyer and the chief architect of the contract that had paved the way for the $4.5 million investment, remembers this not as a disaster at all, but as one of many bumps in the road when the world is trying to develop defences against a new pathogen. "It was just the normal thing in the scientific process. Sometimes good ideas work, and sometimes they don't. This one did not," he said.

As hugely disappointing as this was in the moment for the Australian scientists, it's important to frame the project in the context of the time and the situation facing the world. At this point now, several years later, it could, of course, be seen solely as a mistake. A failure. A bad call. The University of Queensland candidate Covid-19 vaccine did not, after all, make it into full development and did not go on to provide protection for Australia's population, or any other population for that matter. And the $4.5 million was gone. Yet at the time it was one of a clutch of the very fastest start-ups of potential Covid-19 vaccines to get up and running in a portfolio designed to be as broad-reaching as possible in both geography and technology. By the start of June 2020, barely six months after the novel coronavirus had made its unwelcome entry onto the world stage, CEPI had given initial support and funding not just to this Queensland team, but to a total of nine separate scientific and pharmaceutical research teams working on potential vaccines against Covid-19. These were: Germany's Curevac Inc; America's Inovio Pharmaceuticals; Novavax Inc in the United States; Moderna, also in the U.S.; the University of Queensland team in Australia; a research team at the University of Hong Kong; Britain's Oxford University vaccine development team led by Dame Sarah Gilbert at the Jenner Institute; an international research consortium led by France's Institut Pasteur; and Clover Pharmaceuticals in China.

Many of these projects are not well known. In the risky business of research and development, projects and products often fall by the wayside. Thankfully, the most recognisable names are those of companies that ultimately produced an effective Covid-19 vaccine that was subsequently and is still being deployed across the world. Yet, just as it was for Professor Young and his team, the situation for CEPI – which can now claim a catalytic role in seed-funding three highly-effective Covid-19 vaccines that are protecting millions of people across the world from severe illness – could easily have been different. "There's possibly a world in which those early cases didn't sneak out of China," says Richard. "Where China had contained the virus and had somehow managed to put the Covid-19 genie back in the bottle." Then, CEPI – which is funded largely with public money from the coffers of governments around the world – could have been left holding upwards of $30 million worth of commitments for a crisis that hadn't materialised and might never recur. "The question I asked myself in January was 'is that a position I am willing to be in'. And the answer at the end of the day was 'yes'," Richard recalls. "I quickly decided: if I'm wrong, I'll just have to explain why I spent $30 million. I was sure this was a very dangerous virus. I had to take the risk."

So he did. And he didn't stop there. By December 2020, a little less than a year after the novel coronavirus had emerged and begun spreading in people for the first time, CEPI had committed more than $1.1 billion in Covid-19 vaccine research, development and manufacturing agreements. "We thought through each step, how it could be done and came to the conclusion that, yes, there was risk associated with this, but it was a risk that we could take," said Dick Wilder, CEPI's General Counsel.

A study published by researchers at the U.S. National Bureau of Economic Research in Massachusetts in March 2020[13] concluded that the key to a successful research and development (R&D) program is its ability to encourage appropriate risk-taking. The optimum R&D strategy is one that tolerates multiple failures in the pursuit of big rewards. The same study also noted that "appropriate risk-taking is... important because projects with greater uncertainty have a lower probability of bearing fruit, but may also generate more path-breaking innovations if successful."

In a pandemic situation, with the potential for exponential spread and surrounded by that thick, obfuscating fog of war, the level of uncertainty could hardly be higher. As, too, could the potential benefits of a breakthrough innovation hardly be greater. Think of the tens of millions of lives saved by the few successful effective Covid-19 vaccines that emerged from the 340 or so that began development in early 2020. Among those was what became known as the Oxford-AstraZeneca vaccine, devised by Sarah Gilbert and colleagues at Oxford University, originally seed-funded by CEPI and the UK government, and later taken through multiple risky stages of development by the Anglo-Swedish pharmaceutical company AstraZeneca. In an interview in April 2020 while I was reporting on the unfolding coronavirus pandemic for Reuters, Professor Adrian Hill, the director of the Jenner Institute at Oxford University, told me his team had already begun lining up large-scale vaccine manufacturing sites in as many as seven countries across the world. The factories were being readied to churn out an initial requirement of a million doses of a product which at that stage had no name, no supporting evidence from clinical trials, and was at risk of being useless if those upcoming trials showed it didn't work. His colleague Sarah Gilbert, the vaccine's co-developer, later described how much of the project's success was down to researchers, funders, developers, investors, manufacturers and even governments putting in pre-orders and all taking parallel and back-to-back risks as it progressed.[14] This was doing things "at risk" on a pandemic scale.

Infectious disease and public health experts who have, as Richard describes it, spent a lifetime "worrying about pandemics" may not seem likely to be risk-takers. But they are often better equipped than politicians to assess and navigate global public health crises. This is partly, of course, because they do not rely on votes to keep them in position. But it's also because they tend to be more comfortable with uncertainty, more familiar with the fog, and more courageous when it comes to acknowledging a need to change course. "I completely understand the cognitive challenge for politicians and officials – of having to wrap your head around a threat which hasn't manifested, which is invisible, and which is exponential," Richard says. "Unless you're really tuned into that particular kind of threat – which, for historical reasons I was – it's really hard to make sense out of it." The pandemic worrier approach is more akin to risk management than risk-taking, but it's important to understand that managing risk doesn't mean stifling risk-takers' imaginations or starving their appetite for risk. It means being calculating and cognisant of what the risks are, and being aware of what you can reasonably expect to know, and not know, when you take a decision or make an investment.

Besides Richard, one of CEPI's chief risk-takers is Melanie Saville, the coalition's fast-moving Executive Director of Research and Development. Small and slight, with a neat, grey-blonde pixie-cut hairstyle and alert, darting eyes, Melanie can be a tricky woman to pin down. That's not at all because she doesn't want to be helpful, or to talk enthusiastically about viruses, infectious diseases and how to develop vaccines against them. It's because most of the time, she's extremely, overwhelmingly busy *doing* all of those things – and more. Often dressed in a smart dark trouser suit and distinctly not flamboyant or flashy, Melanie inspires a kind of quiet awe among colleagues past and present. Scientists who worked alongside her during her 11 years at the French pharmaceutical firm Sanofi Pasteur describe her dogged and relentless work ethic as remarkable. One of the most important things about deciding what to do in an emerging infectious disease outbreak and when to do it, says Melanie, "is understanding that you're never going to know enough to feel totally confident that what you're doing is right, and understanding that that doesn't matter as long as you're aware of as much as possible." What we saw around the world with the emergence of SAR-CoV-2, she adds, "was a lot of people waiting, hanging back on deciding whether to act before they knew whether this (new virus) was going to be a problem."

Melanie's approach had to be different. She had to be able to take stock, accept that she couldn't be at all sure what the novel coronavirus outbreak was going to turn into, and then decide, in her words, "to go ahead full steam until we know that we don't need to any more." Research and development, the life-blood of medical and scientific progress, is all about taking risks. And scientific progress, as we have seen in the speed and effectiveness of the Covid-19 vaccine development and delivery, is at the heart of getting a handle on viral epidemics. To do this before they gain the momentum needed to spawn a pandemic – in other words to be able to contain a future Disease X before it spirals out of control – we need to be prepared to take risks.

First, though, it's useful to think about the particular scenarios we're exploring here. As Richard would put it, it's important "to understand the kind of movie we're in." There's a special nature inherent in epidemics and pandemics that makes risky decision-making all the more... well, risky. That special nature is the exponential. Infectious disease outbreaks caused by highly contagious viruses can swiftly begin exhibiting exponential, rather than linear, growth. And a big problem is that most human minds – even many of the best and brightest ones – find exponential growth extremely difficult to figure out and deal with. Think of the legendary tale, beloved by maths teachers the world over, about the origin of chess. An Indian emperor, so delighted with the new black and white squared strategy game being presented to him, promised to reward its inventor with whatever he wanted. The chess guru responded that his desires were simple: *"I only wish for this. Give me one grain of rice for the first square of the chessboard, two grains for the next square, four for the next, eight for the next and so on for all 64 squares, with each square having double the number of grains as the square before,"* the story goes. It's a maths problem that still induces wide-eyed wonder among school kids when they figure out, with help, that the emperor would need to put 9.223 quintillion grains of rice on the last square alone – more rice than any country or even the world could produce.

Thankfully, with infectious disease epidemics, we have never reached the last square, or anywhere near it. But the scenario of numbers doubling in a certain (short) time period – when one case becomes two and two become four, then eight, then 16, then 32 and so on – is distinctly different from a scenario of linear growth, where a gradual uptick in the curve over time remains just that – a gradual slope – instead of the reverse ski-jump curve we see with exponential growth. Part of the difficulty in distinguishing one from the other is that both linear and exponential growth start relatively slowly. I remember watching the World Health Organization's infectious disease outbreak specialist Bruce Aylward showing line charts depicting exponential epidemic growth during a media briefing when he returned from a fact-finding mission to China in February 2020. Aylward warned then that "every day we stop to think about this disease and make decisions, should we do it or not, this virus will take advantage and almost double the number of cases." An uncomfortable illustration of this came at the start of the Covid-19 outbreak in Britain, where the government's health agency, Public Health England, issued a statement on 22nd January 2020 saying that in light of rapidly rising case numbers in Wuhan, the Chinese city where the novel coronavirus first emerged, it was raising its risk level for Britons from "very low" to "low." At that point, there were no known or confirmed cases of the infection in the United Kingdom and the official death toll in China was just nine. Three months later, by 22nd April, the viral epidemic of Covid-19 disease in the UK had killed more than 41,000 people.[15] Clearly exponential growth is something we need to be able to comprehend more successfully, and need decision-makers to be able to recognise and communicate more accurately, if we're to understand how to assess and take the risks needed to contain and neutralise a potential pandemic.

The response of political, economic and many public health leaders to looming pandemics is often problematic precisely because of this exponential nature. That's entirely understandable. Heads of ministries, governments and some businesses are expected to be more or less equipped – depending on their level of competence – to deal with a range of crises. To be sure, an economic crisis is very different from a political crisis, which is, in turn, very different from a crisis of conflict or geopolitics. But most situations – even crises – that political leaders have to grapple with are things that unfold in a linear fashion: they develop over time, get worse, get better, take a new turn and so on. In an exponential crisis, by contrast – like the uncontrolled spread of a new viral pathogen – there's a vast lack of information causing confusion, and that is compounded by a lack of understanding of the exponential threat. Elected officials can win praise for having consensus-building skills. Those who deftly assimilate a range of opinions and then plot a course of action that best reconciles the different contending forces within the electorate can find themselves rewarded with the respect of their peers and the votes of their people. However, a less appreciative name of this kind of approach is "groupthink" – something a British parliamentary "lessons learned" report – ironically, led by the former health secretary Jeremy Hunt (we'll meet him again later) – accused the UK government of in October 2021. The report judged that the UK, along with many other countries in Europe and North America "made a serious early error in... not considering a more emphatic and rigorous approach to stopping the spread of the virus, as (was) adopted by many East and South East Asian countries."[16] In an exponential scenario, when politicians and public health officials take a typical crisis-management approach – assimilating information and plotting a path based on linear assumptions – they will, without doubt, be far too slow to make decisions. "They think they are managing risk, but they are misunderstanding the movie they are in," says Richard. Infectious disease crises have unique attributes and unique pitfalls. They demand a bold approach to risk-taking. "The problem... we have at the moment is everyone is afraid of making a mistake – everyone is afraid of the consequence of error. But the greatest error is not to move," the World Health Organization's emergencies programme executive director Mike Ryan noted in December 2021.[17]

Countries that fared relatively well during the early stages of the Covid-19 pandemic did so by identifying a potentially exponential situation, taking responsibility and acting decisively even when there was little knowledge and virtually no certainty. South Korea is widely cited as one of the most successful responders. Despite being one of the first countries hit by the new coronavirus – it reported its first case on 20th January 2020 – and being one of the world's most densely populated large countries, it was able to contain its first wave of Covid-19 infections largely without having to resort to border closures or widescale lockdowns. It did so by immediately conducting comprehensive testing, contact-tracing and quarantining of infected cases. China's approach was more draconian. Yet it was also among the most successful examples of a national response to the emergence of Covid-19. Authorities there did not engage in groupthink. They did not assimilate all the different contending forces and opinions within Chinese society and use that as a basis on which to decide how to move forward. They acted. Risking economic recession and domestic and international consternation, they locked down key cities and regions, imposed strict quarantines and limits on people's movements, and rapidly built vast and expensive hospitals at risk just in case their thousands of beds would be needed for sick patients. Notably, both South Korea and China had recent history with other contagious and deadly coronaviruses – the former with MERS in 2015 and the latter with SARS in 2003. An analysis by researchers at Harvard and at Seoul National University[18] found that when Covid-19 struck in South Korea "the painful memory of MERS inspired an early and aggressive government response" as well as "a willingness among people to wear masks, cooperate with contract tracers, and otherwise listen to public health officials." A report by a World Health Organization-led international team who went to China in February 2020 – just a month into the epidemic – said the regime's "uncompromising and rigorous use of non-pharmaceutical measures to contain transmission of the Covid-19 virus" should provide "vital lessons for the global response."[19] Richard's sum-up analysis is that "they understood the exponential movie they were in and they acted accordingly." And, until a new and highly transmissible variant of Covid-19 known as Omicron began causing a resurgence of coronavirus infections in the Spring of 2022, both South Korea and China had been able to reap the upsides of those risks – freeing their people from restrictions earlier and seeing a swifter rebound in their economies than many other countries in the world.

Countries who fared less well in the initial waves of the Covid-19 pandemic – among them Britain and the United States – and who registered some of the highest death rates in the world in the year before effective vaccines became available, did so often because their governments were seemingly paralysed into inaction. So distracted were they by potential political and economic downsides that they were unable to take risks.

At a time when the then British Prime Minister Boris Johnson should really have known to listen, more than eight months into the national epidemic of the coronavirus and at one of the most intense points of the global pandemic, scientific advisers in a group known as SAGE called on him to take what would, for sure, have been a politically unpopular and risky decision: impose a short lockdown – known as a circuit breaker – to close cafes, bars, restaurants, gyms and other gathering places, to ban household mixing and to move school, college and university learning online. A briefing paper submitted to the government along with their urgings could not have been clearer. "A package of interventions will need to be adopted to prevent this exponential rise in cases," it said. "Not acting now to reduce cases will result in a very large epidemic with catastrophic consequences."

But Johnson did not want to take the risk. An account of the episode given by Johnson's then special adviser Dominic Cummings in *Spike,* a book by Jeremy Farrar and Anjana Ahuja about the UK's response to the pandemic, describes how officials desperately tried to persuade him that a short, sharp, circuit-breaker lockdown would be better for the economy as well as for health. But, Cummings claims, "Johnson basically said, 'I'm not doing it'."[20]

A perhaps prophetic sign of the British government's ability to move – or not – had come several years before in a then secret simulation exercise in October 2016. Codenamed Exercise Cygnus, the war game, or "tabletop exercise" as such projects are sometimes called, involved almost a thousand "players" from central and local government, health service organisations and emergency response planning teams, as well as serving ministers in government at the time. The scenario was an outbreak of a new type of flu, and the simulation was designed to be played over three days. The fictitious virus – dubbed H2N2 or "swan flu" – had emerged in Thailand in June of that year and players were asked to war game how they would handle an already seven-week-old spreading pandemic. In the UK, the fictional national epidemic had already infected up to 50 per cent of the population and claimed the lives of 40,000 people. Media reports about the 2016 exercise leaked out in dribs and drabs over the following years as well as in the early stages of the real pandemic in 2020 after the government declined to publish them until forced to by freedom of information requests. According to sources who took part quoted in those reports, the then UK Secretary of State for Health, Jeremy Hunt, downed tools and "refused to play" when the simulation required him to make a notional, politically risky decision: whether to remove critically ill patients from ventilators and let them die so as to free up space for less severely ill patients who would be more likely to benefit – in other words, more likely to live – as a result of getting ventilator treatment. "The hospitals were full and Hunt was asked to make the call as part of the exercise. But, instead of doing so, he basically said 'I'm not playing anymore'," Britain's *Daily Telegraph* newspaper quoted a source from inside the simulation as saying.

Kate Bingham, who led one of the UK's more successful aspects of the Covid-19 response – the swift, extensive and expensive rollout of a nationwide vaccination programme – later described the British government as being dominated by process rather than outcome, causing delay and inertia. "There is an obsessive fear of personal error and criticism, a culture of groupthink and risk aversion that stifles initiative and encourages foot-dragging," she said in a lecture at Oxford University in November 2021. "Government must be braver."[21]

As Richard found out from his experience at the White House during the 2009 swine flu pandemic, a tendency among political and public health leaders to play safe is nothing new. A preference for risk aversion can span decades and the political spectrum. In a handwritten personal journal Richard kept during that 2009 pandemic year, he noted that while Mexico had taken a bold policy plunge and closed schools to try and stem the spread of the H1N1 virus, the United States' pandemic response appeared to be driven more by a desire to avoid taking any such policy risks.

"Carter and I argued as hard as we could," Richard wrote. "It was an argument about the precautionary principle v(ersus) the scientist's desire not to make a mistake, coupled with risk aversion that was all too characteristic of public health bureaucrats."

Roll forward 21 years or so, and little had changed among decision makers. By the start of March 2020, the U.S. had already been seeing a growing coronavirus epidemic for at least a month, but then President Donald Trump repeatedly delayed, ignored or dismissed the need to make decisions – including on closing schools. Writing to Richard on the pandemic-watching Wolverines email chain that had become an invaluable source of what Richard called "sensemaking" during this and previous global health crises, Carter Mecher noted on 5th March that "just about every country with community transmission is closing schools, including the best practice places like Singapore and Hong Kong," but that in the United States, "our political leaders and public health leaders are flinching." "The political leaders, emergency managers and even public health leaders understand that they must take very disruptive actions early during a hurricane (evacuate while the sun is shining and a mild breeze)," Carter went on. "Not sure why they cannot apply that same mindset to disease outbreaks."

What's most likely to be behind the flinching is a fear of things going wrong. Officials and leaders who make bold moves early on in a growing infectious disease epidemic can secure great rewards in terms of deaths averted or even pandemics thwarted. But failures can hit hard.

One such scenario is seared on the minds of many American public health officials in particular, and is known to be one Obama re-analysed and deliberated over during the pandemic of 2009. It was a swine flu outbreak back in 1976 which has gone down in history, arguably not entirely fairly, as a "fiasco." It began at a military training base in New Jersey called Fort Dix. In February 1976, a handful of soldiers at the U.S. army base came down with flu. Tests showed while some had the expected seasonal flu virus, others had been infected with a new variant of an H1N1 swine flu. Since none of the soldiers had been in contact with pigs, health officials moved swiftly to test hundreds more of the recruits to see if what they feared – a new, human-adapted H1N1 swine flu spreading person to person – was actually happening. The tests showed the contagion had spread to more than 200 soldiers. Still keenly aware of major deadly flu outbreaks in 1957 and 1968, and yet more spooked by the H1N1 tag that harked back to the 1918 Spanish Flu that had killed 50 million people worldwide, U.S. health officials worried that the new flu variant posed a high risk of a pandemic. They decided to act fast.

Reflecting on the episode, the 90-year old vaccine expert, doctor and professor Stanley Ploktin, noted that the then director of the CDC, David Sencer, acted swiftly and dramatically. Sencer called for a nationwide vaccination programme to nip in the bud what he perceived as a major epidemic, even pandemic, threat. Within weeks, and working in tandem with Sencer, the then President Gerald Ford launched a $137 million project to develop and produce a H1N1-targeted flu vaccine before the autumn. Two million doses were ordered. The downside to this ambitious plan was that not only did the epidemic never take off, but the vaccine itself was found to cause a rare but severe and sometimes paralysing side effect known as Guillain-Barré Syndrome in a small percentage of its recipients. By the following year, the outbreak had petered out, the public's confidence in vaccinations had been severely damaged by the cases of rare side effects, and Sencer, after a record 11 years as CDC director, had been sacked for overreacting.

"It was a chancey thing," said Plotkin, who, as the godfather of vaccinology and inventor of the Rubella vaccine, is intimately familiar with all aspects of risk that the field presents. "The epidemic never occurred and the whole thing cost the job of the head of the CDC. From my point of view, what happened was a good thing, because they reacted to a potential epidemic in a proper way. Unfortunately, that's often not the way the world sees it."

Part of preparing to take risks, then, is preparing to enable others to take them. That probably also means accepting that chastising them later if one or two of their decisions turn out to be wrong is going to be counterproductive. Just as one way not to foster innovation and decisive action is to throw decisions to a large group of people, all of whom are going to think of different ways to point out obstacles or slow things down – groupthink – another is to unforgivingly analyse prior risk-taking moves and call out only the downsides. When "lessons learned" exercises do this, and then put in place systems and processes designed to reduce the risk of failure, they stifle risk-taking and limit chances of future success. Shortening the rope on the downside also means reigning in the potential upsides, and ultimately starves innovators of the space they need to break new ground that could save millions of lives.

When it comes to preventing future pandemics, risk-taking is a little different. Unlike recognising and encouraging the types of pandemic-scale decisions taken, or not taken, by the scientists, politicians and public health officials described above, identifying and characterising pandemic-preventing risks is relatively easier. Mostly, those risks come down to money. Investing at the right time, at the right level, in the right defences.

Preparing the world to be able to neutralise future looming pandemic threats is based in part on building a vast viral surveillance and early-warning system, a vast base of scientific, technical and manufacturing know-how, and a vast library of prototype vaccines that could be swiftly adapted and used against any one of a plethora of 300 or more viruses that might one day mutate in a way that enables it to spark a human contagion. When United Nations Secretary General Antonio Guterres spoke at the March 2022 Global Pandemic Preparedness Summit in London he put it like this: "The world needs a global pandemic preparedness system that addresses problems before they turn into catastrophes."

There's no running away from the fact that, to be truly comprehensive and effective, such a system will be expensive. Putting state-of-the-art viral surveillance and next generation genetic sequencing technology into all the world's spillover hotspots won't come cheap. USAID's decade-long PREDICT project cost $238 million and only covered a tiny, tiny fraction of the surveillance scope needed to spot the next Disease X earlier, and did it for only 10 years. And as we've found out during Covid, developing vaccines is a risky business that can cost billions of dollars with a high chance of failure.

But the question then becomes how much money are we willing to risk spending each year to be able not to have another disaster like Covid? Adding a little context as a frame to that question, here are a few numbers from the Covid pandemic: by the end of March 2022, estimates of global cumulative excess deaths due to SARS-CoV-2 infections ranged from 14 million to 24 million people – almost matching the death toll from World War I. The expected global economic cost of Covid by the end of 2025 is put at $28 trillion.[22]

In part to test out how much wealthy governments and global financing agencies such as the World Bank had begun to develop a more realistic appetite for financial risk when a potential pandemic is imminent, Richard took part in another simulation exercise in February 2022. This was a time when the world was very much still in the midst of a real pandemic, Covid-19, and very much still suffering its devastating human and financial costs. This war game, conducted via Zoom with representatives from various G7 and G20 countries brought in to put their respective national positions, was co-led by the German government and played out as a side event to that year's Munich Security Conference. It was what is known as a "closed-door event" and was for government officials and technical advisors only. Its title was *Every Day Counts: A Pandemic Vaccine*

Exercise. To start off with, the scene was set with news reports about an outbreak of a fictional viral disease called "Clade X" spreading rapidly from two separate starter clusters of infections – one in Europe and one in South America. The fictional disease, characterised by fever, cough, confusion and, in many cases, coma, respiratory failure and death, had already killed an estimated 74 people and infected at least 300 in the two separate outbreaks. The first death, according to the simulation, had occurred within days, and there are clusters of infections reported in family groups and among healthcare workers. Preliminary assessments from the start of the outbreak in Europe suggest that, on average, a Clade-X-infected person in the fictional scenario transmitted the virus to between two and three other people. This made Clade X about as contagious as seasonal flu and as the 2003 SARS virus, the scenario proposed, but less contagious than the Delta or Omicron variants of the Covid-19-causing SARS-CoV-2 virus.

Asked to imagine themselves called to a meeting several weeks into this situation, participants were invited to say whether G7 and G20 countries would agree at this point to step up and put a joint fund of $20 billion on the table for at-risk research, development, manufacture and purchase of six potential vaccines against the newly emerging pathogen. Alternatively, the participants were asked, should they do what was done for Covid-19 and mostly provide funding individually without a commitment to sharing? Their answer – even at a time when the unprepared-for Covid-19 pandemic was draining their real governments' coffers drier by the minute to the collective tune of trillions of dollars – was equivocal. Only one or two participants said their countries would be full-scale behind the creation of a $20 billion joint fund, and several others said their primary focus would be the same as it had been during the Covid-19 pandemic: spending millions of dollars on securing bilateral deals for potential vaccines for their own citizens, even if those vaccines might not ever materialise, or might never be required if the fictional Clade X turned out to be less contagious and deadly than feared. Making a decision to make money available for a multilateral at-risk fund was seen by several players as a lower-priority, optional-extra risk.

Watching this exercise as an observer bound by strict rules on naming no names and giving no attributions, I witnessed what the world had not learned. A call to put a measly one billion dollars each at risk, from countries that could easily afford it, to shore up vital global defences against a potentially pandemic-causing viral threat, was still seen as too much.

It was something that clearly struck several of the simulation's other participant-observers too. One promptly suggested something that Richard and other pandemic experts have increasingly been arguing for – that governments need to think about spending on preventing pandemics the way they think about investments into national and regional defence and security. Imagine if governments started spending on research into new types of weapons, figuring out new conflict strategies, or building new manufacturing plants to produce munitions and other tools of war only after an invasion had started. "That would be a completely unworkable model," Richard says. As is it a similarly unworkable model for pandemic preparedness. In essence, the preparedness surveillance and science we need to advance is a defence system that, if it is created in a comprehensive and robust way, can take the threat of pandemics like Covid-19, or worse, off the table. Viewed that way, paying to prepare against pandemics looks like a risk governments and their people should be more than willing to take.

4: PREPARE to Share

Because I was born in Sokoto, northern Nigeria, in late 1969, I have a scar high up near the shoulder on my left arm that my kids don't have, and will probably never need to have. That little skewed oval of puckered, shiny skin marks me as among the last direct recipients of a medical defence tool that brought about one of humanity's most extraordinary feats of global health collaboration: the worldwide vaccination programme that eradicated smallpox. With no cases to remind us, it's easy to forget the scale and ferocity of smallpox. It was – is – a truly horrific disease. The infection is caused by the variola virus, which is most often breathed in by its victims. Typically, it starts with a high fever, muscle aches, headaches and vomiting. A few days later, a rash begins to appear on the tongue, mouth and throat in the form of red spots and sores. Within a day or so, the rash spreads in little bumps to the skin of the face, and then on to the arms, legs, torso and all over the body, including the palms of the hands and the soles of the feet. Only then do the disease's signature pus-filled pustules feature, gradually bursting and scabbing up over the following days. Those who died of smallpox usually did so within one to two weeks.

In the 20th Century alone, the disease is estimated to have killed between 300 million and 500 million people. Among those it didn't kill, it left many either blind or horribly disfigured, or both.

A pioneering foot soldier in smallpox's eradication was William Foege, an American infectious disease specialist who, inspired by his uncle's work as a missionary in New Guinea, decided in December 1966 to sign up with the U.S. Centers for Disease Control and Prevention (CDC) to work on tackling smallpox in Nigeria. The original thinking among infectious disease experts at the time had been that any decent attempt to wipe the disease out globally would require mass vaccination of at least 80 per cent of the population in any country still seeing cases of infection. Yet when Foege heard from missionaries communicating via short-wave radio that there were smallpox outbreaks in several eastern Nigerian villages, he knew he did not have enough doses of vaccine to hand to launch a country-wide mass vaccination campaign. Instead, he travelled to the affected villages and vaccinated only the people who lived there and in one or two neighbouring villages that had regular contact with them. His strategy, which become known as "ring vaccination" for its ability to create a ring of protection around cases of a disease to stop them spreading outwards, proved highly effective and was subsequently adopted across the rest of the smallpox fight. It was just over 10 years later, on 8th May 1980, that the World Health Assembly endorsed the historic WHO declaration "that the world and all its peoples have won freedom from smallpox." It was the first and, so far, the only human disease ever to have been eradicated.

Following on from his work in Africa, Foege rose up the CDC ladder and went on to replace the unfortunate David Sencer as the agency's director in 1977, holding the post until 1983. In a memoir about the 10-year eradication campaign, *House on Fire,* Foege described the plan's ultimate success as proof that "humanity does not have to live in a world of plagues, disastrous governments, conflict, and uncontrolled health risks."[23] This was because, he went on, "the coordinated action of a group of dedicated people can plan for and bring about a better future."

The critical trigger for the push to wipe this lethal and maiming disease from the face of the planet was not the disease's sudden arrival:

after all, smallpox, which had a terrifying death rate of up to 30 per cent, had existed in humans for thousands of years. Nor was the decision to try to eradicate smallpox triggered by a stunning medical breakthrough: a vaccine against the disease had been around since the British scientist Edward Jenner invented it in the late eighteenth century. The trigger for a concerted effort to deploy that vaccine to try to end the decades of suffering and death was a decision – even at the height of the Cold War – by the Soviet Union and the United States to put serious and money and effort into working together to fight a common enemy.

It was a massive undertaking. Smallpox was at that time infecting more than 15 million people around the world and killing as many as two million each year. But with huge donations of vaccine from both sides – 25 million doses a year pledged by the Soviet Union and 50 million a year by the Americans – what ultimately made the campaign a success was not just scientific progress, but diplomacy, cooperation, solidarity and collaboration. Put simply – it was about sharing the work, sharing the responsibility, sharing the knowledge, sharing the technology and the tools, and ultimately, sharing the benefits.

These days, it's the World Health Organization that is widely credited with the success of the smallpox eradication campaign. Yet the idea to go for it and try to wipe the horrific disease from the face of the planet came first from the Soviet Union. Another account of the programme written by the American scientist at its helm, Donald A Henderson, in 1998, noted that even though the USSR had at that time already eliminated smallpox from within its borders, it was "regularly besieged by smallpox importations from abroad" and hence forced to keep up an expensive nationwide vaccination campaign to keep it out.[24] Similarly, smallpox was already eliminated in North America and Europe, although there were still cases occurring in South America, Asia and Africa. When the WHO had asked Henderson in 1966 to take the lead on the campaign, he calculated that having an American at the head of a project which the Soviets had championed might upset the USSR. Not keen to start off on a bad footing, he invited the Soviet Union's then Deputy Health Minister, a smallpox eradication enthusiast called Dimitry Venediktov, to co-lead it with him. "Why not join forces to attack the problem on a global scale," Henderson wrote.

It's heartening to think that even at a time when Washington and Moscow were pointing missiles at each other as arch Cold-War enemies – sadly much the same as they are again now – the coordinated action of a group of dedicated people was able to bring about a world first in beating a disease. And while there are many differences between smallpox and the types of infectious diseases that have pandemic potential, the lessons of the global effort to fight for and achieve its eradication are clear. Galvanising a scared and fractious international community to defend itself against a common threat is not just possible, it's essential. To be able to get ahead of a potential pandemic, we need to prepare to share.

In 2020, Richard was acutely aware of this need – almost as swiftly as he was aware that the novel coronavirus spreading in Wuhan would soon engulf the world in a deadly pandemic. On 23rd January of that year, the same day he had announced at a press conference in Davos the agreements to invest in Moderna's, Inovio's and the University of Queensland's Covid vaccine development projects, he sat down for early evening drinks and a shared plate of nachos in the bar of the Hard Rock Hotel in the Swiss mountain resort. His drinks companions were Dr Cynthia de Vivo, a specialist physician, and her husband, Dr Seth Berkley, a curly-blond-haired larger-than-life American doctor who had in 2011 moved from a not-for-profit, public-private partnership he'd founded to accelerate scientific research into potential HIV vaccines to become chief executive of GAVI, the Global Alliance for Vaccines and Immunization. The GAVI Alliance was set up in 2000 and uses money from philanthropic, private and government donors to negotiate down vaccine prices for developing nations and then bulk-buy and deliver them to countries whose populations need them most.

A bit like it did for the likes of Dr Tedros Adhanom Ghebreyesus or Dr Anthony Fauci, the Covid-19 pandemic has made Dr Seth Berkley something of a household name. He is known as someone who is always on the go, flying all over the world taking on challenges, trying to solve problems and encouraging others to do the same. "I can be a little hyperactive," he admits. He didn't even stop when he made his marriage proposal to his now wife Cynthia, choosing instead to pose the question while on a plane ride to South Africa. He was on *Fortune* magazine's list of the world's 50 greatest leaders in 2021, among several other global health and vaccine champions including the pioneering scientists behind the mRNA technology which helped speed Covid-19 vaccines into development. He had previously also been featured on the cover of *Newsweek*, recognised by *TIME* magazine as one of the "100 Most Influential People in the World," and picked by *Wired* magazine as one of "The Wired 25 – a salute to dreamers, inventors, mavericks and leaders."

My own first encounter with Seth was during a global health reporting trip to Ghana in West Africa a decade ago, in 2012. The GAVI team, with Seth at its helm, were there to support the Ghanian government's introduction of two new childhood vaccines – against pneumococcus, which can cause meningitis, sepsis and pneumonia, and against rotavirus, which can cause severe diarrhoea – into its national immunisation programmes. And I was there, along with correspondents from other media outlets around the world, to report on the historic double new vaccine rollout, as well as to conduct interviews and research for a feature I was writing about cancer in Africa. Chatting with Seth in an interview about his career, he told me how he'd moved from a childhood in New York where he talked his way into helping out at a retail chemistry supply store to get closer to science, to working in a ghetto clinic in Mississippi, to studying tropical medicine in Brazil. "I love science and I believe in it. I have a faith that science can solve problems and make the world a better place," he said. He also told of how, as a young epidemiologist working for the U.S. State Department, he'd witnessed the devastation caused when measles swept through refugee camps in Sudan during the 1985 famine. Measles is history's most infectious human disease. It is far more contagious than any strain of flu, pandemic or otherwise, that we have seen so far. It spreads far further and far faster than Ebola or Covid-19. The virus that causes it hangs in the air for more than an hour after any infected person has moved on. So in a vulnerable, unprotected population, each infected person can pass it to at least 12, and sometimes as many as 18 others.

"You'd see little shallow graves, lined up, one after the other – babies," Seth said as he described what was left in the Sudanese camps after a deadly wave of measles. It was sights like those child-sized graves, he said, that had helped him evolve from a curious, adventure-seeking young doctor into a man with "a mission to vaccinate every kid on earth."

That trip to Ghana when I met Seth in person brought me face to face, too, with the reality of unequal access to life-saving defences against infectious diseases. Face to face with the reality of what not sharing means for the have-nots. Visiting the 74-bed Princess Marie Louise Children's Hospital in the Ghanian capital city of Accra on a sun-soaked day at the end of April, I came across a sad scene in a dim, shadowed ward on the facility's ground floor. The infectious disease enemy in this case was not measles, but pneumonia – a disease that in many childhood cases can be prevented with a vaccine. The vaccine had at the time been available in the United States since 2000, in Belgium since 2004 and in Britain since 2006. But – until Seth and GAVI intervened to bulk buy it and make it available at an affordable price – it had not been accessible for many poorer countries like Ghana.

On what looked, by comparison with its inhabitant, like a huge hospital bed made up of a pillow, white sheets and a mattress on a grey metal frame lay a tiny boy. His name was Isiah Anane, and at eight months old, his first two baby teeth were just visible in his gummy mouth. His little body – limp limbs, a bare tummy and chest emerging from a loosely-fitted nappy – was heaving as he struggled for breath. His father wailed and wept at his bedside, offering up prayer after prayer for his first-born to be saved from the infection flooding his lungs and the fever sending his body into seizures. Even with a lifeline tube leading from a battered blue and white oxygen tank on the floor at his bedside up and over the bedrail and into his nose, Isiah's lungs were barely getting enough oxygen into his blood to sustain him.

Whispering so as not to interrupt Isiah's father's pleas, I asked the emergency room's specialist paediatrician, Dr Margaret Neizer, about the case. The baby had been brought in at dawn, she said, with a dangerously high temperature and already very weak. "This is why we need the vaccines," she told me – to stop these kinds of preventable infections taking hold in the first place.

Late that night, after I'd gone back to the hotel I was staying at in Accra, Dr Neizer called GAVI's media support officer to report what I had been dreading. Isiah had died that evening. He'd become one of the thousands of babies in Ghana and beyond at that time whose lives could have been saved by a rich-world-manufactured vaccine, but were not.

As Seth, Richard and Cynthia sat down to drinks in the bar of Davos' Hard Rock Hotel on that January 2020 evening, they were grappling with a new infectious enemy. Barely a handful of cases of the disease caused by the novel coronavirus had been reported outside China, yet these seasoned infectious disease warriors were already thinking globally. To be true to the pandemic mantra – that no-one is safe until everyone is safe – they knew that defending against this disease if it went global, as they already in January 2020 suspected it might, would mean sharing any new knowledge, technology and weapons that could thwart its progress as widely and as swiftly as possible. They were mindful, too, of what had happened during the 2009 H1N1 swine flu pandemic, when pharmaceutical companies based mainly in North America and Europe developed a vaccine against the pandemic flu strain in just seven months, but rich countries bought up almost all the supplies. How could the world ensure any vaccines developed against this new disease would get to everyone at risk, everywhere? After seeing how the rich world in general, and the United States in particular, had responded to the 2009 H1N1 swine flu pandemic, the experts were focused on how to avoid history repeating itself. "We asked the question, what happened in swine flu? And what happened... was that, despite the fact that there were eventually vaccines made, they got bought up by wealthy countries and didn't make it into the developing world," says Seth. That had meant that in 2009, the World Health Organization was only able to muster around 78 million doses of swine flu vaccines to provide to just 77 countries. And even then, those 77 recipient countries only got them many months after wealthy countries had had first dibs. In some instances, they got them also only as the pandemic had already begun to peter out. Given that experience, and with the staunch nationalist Donald Trump in the White House at the time, Richard describes his expectations of any U.S.-led campaign for a multilateral approach as "practically zero." "We knew at the time that unless we tried to do something, all of the rich countries were going to buy up all of the vaccines. So the question was how can we design a mechanism that possibly offsets that?"

A couple of months and many hours of thinking and planning later, Richard drafted what he called a "white paper," entitled: "A proposal to establish a globally fair allocation system for Covid-19 vaccines."[25] This was the first written pitch for what would eventually become COVAX – the world's first ever fair allocation system for vaccines to be created in the midst of a global public health crisis. The proposal paper described an end-to-end project that would stretch all the way from the development of new vaccines against the novel disease to the delivery of those vaccines to people across the world. And it would start simultaneously in rich, middle-income, and poor countries with people at highest risk – the elderly, the sick, and the healthcare workers on the pandemic's front lines. "We must avoid any scenario in which high-income countries monopolize the global supply," it warned. Allowing precious new Covid-19 vaccines to be snatched up by a small number of countries would be a recipe for disaster and would ensure the disease would continue to circulate and threaten the whole world even well after a means of stopping it had been developed.

The white paper's central idea was that wealthy and middle-income countries would pool funds into the allocation system to put into the speedy development, manufacture and then bulk-buying of Covid-19 vaccines. The doses would then be distributed equitably to the highest risk populations in poor, middle-income and rich nations as they became available. This way, those most exposed to the disease and most vulnerable to severe or even fatal infections would be protected at the same time, wherever they were in the world. The project could be established in a fairly straightforward manner, the white paper said, as long as there was sufficient political will and public sector financing. "It was an urgent effort, recognising what would unfold if we didn't try, to create a mechanism to offset the inequities that we knew were going to emerge," says Richard.

The mechanism was launched in June 2020 by the WHO together with CEPI, GAVI and the United Nations Children's Fund UNICEF. Designed to be part buyers' club and part insurance policy, its ambitious aim for 2021 was to bring two billion doses of Covid-19 vaccines – vaccines which at that point had yet even to be developed or proven effective – to the countries and their people who would otherwise have little hope of getting hold of them. The project was called COVAX, adding to the plethora of acronyms scattered through the language of global health. This one stands for the Covid-19 Vaccines Global Access Facility.

Despite being built entirely on the fly in the midst of what turned into the most devastating global health crisis for 100 years, COVAX has done astonishing things. Ghana's Kotoka International Airport in Accra was the scene for the arrival of the first 600,000 COVAX-procured doses of the Oxford-AstraZeneca Covid-19 vaccine on 24th February 2021. It had been qualified and authorised by the WHO as proven safe and effective during months of development and clinical trials. The doses were delivered just shy of a year since the World Health Organization had begun describing the outbreak of the novel coronavirus as a pandemic, and less than three months after Margaret Keenan became the first person in the world to get a fully licensed Covid-19 vaccine. Such a short time lag between wealthy nations and poor nations beginning to get supplies of newly-developed vaccines had never been seen before. The smallpox vaccine took decades. And, for Ghana as one example, there'd been a delay of at least six years for access to the types of pneumococcal vaccines that could have saved baby Isiah's life. With Covid-19 vaccines, the time lag was less than 80 days.

Yet COVAX also fell painfully short of its sharing goals. Self-interest and vaccine nationalism on a scale that proved highly destructive thwarted early hopes that the project would rapidly deliver large numbers of new Covid-19 vaccines to poor countries. The uncomfortable but unavoidable truth is that people are fundamentally selfish. And governments reflect the selfishness of their citizens. An analysis by the anti-poverty group, the ONE Campaign, of supply deals for Covid-19 vaccines in February 2021 found that rich countries were at that point on course to have over a billion more doses of Covid-19 vaccines than they needed, leaving poorer nations scrambling for leftover supplies. The United States, the European Union, Britain, Australia, Canada and Japan had already done deals to secure more than three billion doses of the new vaccines, vastly more than the around 2.06 billion they needed to give their entire populations two doses. Britain alone ordered five times as many doses as it would need – hedging its bets by securing deals with at least seven separate developers just in case one or several of their still-experimental products failed in trials. Meanwhile, poorer countries could do little more than stand by and watch as rich nation bilateral vaccine deal after rich nation bilateral vaccine deal hit the headlines. As had happened in pandemics gone by, the haves were able to grab and hoard, and the have-nots were left behind. By June 2021, six months after several proven, licensed, safe and effective Covid-19 vaccines had come to market, the WHO reported that the world's poorer countries had received less than one per cent of the doses administered globally. Vaccine nationalism had taken over.

By the middle of September 2022, COVAX had delivered 1.7 billion doses across 146 countries. This was an amazing achievement in the context of history, but still far, far short of COVAX's aim to get two billion doses shared equitably across the world by the end of 2021. Equity remained a distant hope. As of March 2022, when many people in the wealthy global north appeared to be putting the pandemic behind them, with their thinking about vaccines focused on enviable decisions about whether or not to go for a third or even fourth booster dose, more than three billion people – almost half the world's population – had not yet received a single dose of a Covid-19 vaccine. Speaking at a Global Pandemic Preparedness Summit in London that month, Nigeria's Health Minister Dr Osagie Ehanire put it like this: "Efforts to counter the assault of Covid-19 on humanity seem to have been for the benefit only of a part of humanity." Nigeria experienced its highest peaks of Covid-19 deaths and new cases in the autumn and winter of 2021 – a time when 98 per cent of its people had not had access to the protective vaccines that had been administered to billions across the rich world. By the time the United Kingdom had, in September 2021, vaccinated 71 per cent of its population against Covid-19, Nigeria had managed to get vaccines to only two per cent of its people. At the same point, the United States had achieved Covid-19 vaccine coverage of almost two-thirds of its vast population, while Ghana had vaccinated barely three per cent of its people.

The effects of vaccine nationalism are both immediate and acute, and profound and far-reaching. The price of inequity for the have-not nations is paid by their people. Populations in low and middle-income countries have continued to suffer higher rates of death, hospitalisation, long-term complications and economic hardship in the Covid-19 pandemic than vaccine-protected nations, whose death rates from the disease began to plummet with the arrival of vaccines. But in monopolising and hoarding supplies of vaccine, the haves were creating more than just a humanitarian crisis for their poorer neighbours. They were paying a high price, too. A study by the International Chamber of Commerce in January 2021 found that fully vaccinating the populations of rich countries while neglecting poor ones could cost rich countries as much as $4.5 trillion in lost economic activity.

More importantly, there were the variants. A clear but perhaps not directly intended consequence of the ugly vaccine nationalism that left billions of people in poor countries unprotected was that – as Nigeria's case count data showed – SARS-CoV-2 virus was able to circulate unhindered in largely unvaccinated populations across much of the world. This, without doubt, created a perfect viral storm environment. There were ever more opportunities for the virus to pick up new mutations, including ones that could make it more easily transmissible and more easily able to evade any new vaccine-induced immunity. In short, vaccine nationalism fuelled the evolution of mutant variants – viruses that went on to cause new waves of infections across the whole world.

In the case of Omicron, which has since been dubbed a "super-variant" for its highly infectious nature, it started with an uptick of cases in Gauteng Province in South Africa in mid-November 2021. Scientists who had been regularly sampling and analysing cases of Covid-19 infection reported being shocked at finding that a new version of the virus was behind hundreds of new cases in the area. Sequencing the genomes of samples, they found that many of them had a large number of mutations.

More troubling still was that many of those changes were on the infamous spike protein that the SARS-CoV-2 virus uses to enter human cells. This variant had the potential to spread like wildfire, like no coronavirus variant ever seen before.

Leading the South African team was a long-haired and brilliant young Brazilian-Portuguese-South African virologist and bioinformatics professor called Tulio de Oliveira. Interviewed afterwards about the moment he realised he had found something alarming, de Oliveira described how his team had acted quickly and decisively, "almost like a commando SWAT (Special Weapons And Tactics) team." "I just sent a WhatsApp to everyone saying 'guys, I think that's a new variant'," he recalled in an interview with *Spotlight*, an online South African public health news publication. "(Then) you look to each other, and everyone knows exactly what to do. 'Okay. You go there, you do that'. Everyone is very well trained." Immediately aware of the need to alert the world by doing the right thing with their findings, de Oliveira's team duly shared their data on GISAID, the largest database of viral genome sequences in the world.[26] There, they soon found that scientists in Botswana and Hong Kong had also reported cases with the same worrying genetic mutations. In response, the WHO issued a statement thanking the scientists – in particular Alan Tsang and colleagues in Hong Kong, Sikhulile Moyo and colleagues in Botswana, and Tulio de Oliveira and colleagues in South Africa – as well as the wider GISAID community "for rapidly sharing their genomes to detect this new variant of the pandemic coronavirus" and in doing so making a "a remarkable contribution to global health security." Yet de Oliveira, who later picked up a *TIME* magazine 100 Most Influential People award, was at the time dismayed and horrified at the extreme backlash against South Africa that was unleashed after his team and the government went public so swiftly with the news. Within hours, the 27-member European Union, Britain, the United States, Canada and others imposed or reimposed travel bans to and from not only South Africa, but also many other countries in southern Africa including Botswana, Namibia and Zimbabwe. "If the world keeps punishing Africa for the discovery of Omicron... who will share early data again?" de Oliveira tweeted in early December. "Countries had blocked

Africa and caused massive economic costs. (Now) instead of helping, they are hoarding more vaccines for boosters."

Predictably, given the evidence from all the previous waves of Covid-19 infection that had swept across the world, travel bans and border closures did not stop the wave of Omicron. Barely 10 weeks after the variant was first identified, almost 90 million new cases of Covid-19 caused by the Omicron variant had been reported to the WHO. This was a higher case count than was reported in the whole of 2020, the first full year of the coronavirus pandemic.

The failure of wealthy countries to share life-saving doses of vaccine and to recognise the global and the self-harm caused by vaccine nationalism filled Richard with sorrow – even though he had expected them both. In our morning talks during this time, he referred often to the unbearable pain inflected by governments' inequitable response. Saddened, he worried about how the world had allowed the Covid pandemic to perpetuate in such a way that put everyone, everywhere still at risk. More than two years on from the initial emergence of the novel coronavirus, an uncontrolled virus was still circulating widely and freely enough to be able to adapt and mutate in ways we could not predict or control. "Nobody can possibly be satisfied with these outcomes," he told me. "And part of the reason for them is that we were slow, and part of it is because we haven't shared." The emergence of the Omicron variant of Covid-19 in mid-November came just a few weeks before he was due to speak at a special session of the World Health Assembly to be hosted from the WHO's Geneva headquarters at the end of that month. The meeting, involving all 194 member states of the WHO and hundreds of global health experts from all over the world, was only the second time in history a Special Session had been called. (The first such event had been held in November 2006 after the sudden death of the WHO's then Director General Lee Jong-wook.) Depressingly, the new variant's scary appearance had fulfilled precisely the predictions of the many infectious disease and global health specialists who had warned that allowing the virus to circulate in areas where access to vaccine was limited would only speed its evolution. Having only been able to access enough Covid-19 vaccine doses to vaccinate less than a quarter of their populations, Botswana and South Africa had become fertile ground for mutants to evolve. "The virus is a ruthless opportunist," Richard told his virtual audience. "And the inequity that has characterized the global response has now come home to roost." Omicron and its many mutant SARS-CoV-2 variant cousins are and will continue to be the real-world manifestation of how vaccine nationalism can ensure that "no-one is safe until everyone is safe." And, as they continued to circulate in unvaccinated populations more than two years into the pandemic, they were also another prompt for leaders from low and middle-income countries to stand up and shout "Enough!"

One of those shouting the loudest was Dr Ayoade Alakija, a Nigerian medical doctor and specialist in infectious diseases, public health and epidemiology. In December 2021, she was picked by the World Health Organization to take charge of accelerating equitable access to the newly developed tests, treatments and vaccines for Covid-19 that were then proliferating in the wealthy world. When Ayoade walks into a room, people generally take notice. And just in case they don't, she's already got a preplanned strategy to make them. "If they don't give you a seat at the table, pull up a chair," she says. "And if they don't make space, then get on the table." Tall, striking, confident and often adorned in eye-catching jewellery and fabrics rich with colour and dynamism, she goes by the nickname Yodi and describes herself as "Agitator Number One for vaccine equity."

When I met Yodi in March 2022, she was definitely agitated. We were in London, on the top floor of the iconic Science Museum with a view out across the capital's skyline. This was the venue for the Global Pandemic Preparedness Summit hosted by CEPI and the UK government. Yodi had been invited to speak on a panel that also included Dr David Heymann, an American epidemiologist who had been executive director of communicable diseases at the WHO during the time of the SARS (SARS One) outbreak in 2003, and Sir Jeremy Farrar, who, like Richard and others, had already been warning that a lack of leadership in wealthy nations with the means to share defences was prolonging the pandemic for everyone. Yodi, who splits her time mostly between Abuja and London, had been in Britain during the first national Covid lockdown imposed by the UK government in March 2020. Almost two years on, as she came back to London for the Global Pandemic Preparedness Summit, the then British Prime Minister Boris Johnson's new mantra was that the UK population should be thankful for what he called "our hugely successful vaccination programme" and was now "in the strongest possible position to learn how to live with Covid." Barely able to contain her anger at this myopic, privileged, "I'm-alright-Jack" view of a pandemic that was still raging virtually unchallenged in much of the Global South, Yodi reminded the audience that while they had all been protected with the Covid-19 vaccines developed using cutting-edge technology, millions of other people around the world were still dying preventable deaths. "Here (in London) everybody's walking around like: 'Oh, wow, this is fantastic. It's all over – Covid's over!,' But no, no it's not," she said. "It depends on where you're sitting in the world. Those of us in this room have been double vaccinated, or triple vaccinated, and we all had access to a rapid test as we walked into the building, so we can shake hands, or hug and it's all fine. But many other parts of the world are still breeding grounds for super-variants." And, as we have seen time and time again, super-variants have even less respect for borders or boundaries than do their less super cousins. A system where sharing is either absent or comes only as an afterthought once nationalistic appetites have been sated and even overwhelmed with more supplies than can possibly be used, is one that fails humanity everywhere. No one is safe until everyone is safe.

To be sure, just as the smallpox eradication programme did at the height of the Cold War in the 1960s and 1970s, the Covid-19 pandemic has also thrown up great feats of collaboration. The super-fast new coronavirus vaccines themselves were the product of vast international scientific, manufacturing, supply and delivery efforts that could never have been achieved in one company or one country. Take the Pfizer-BioNTech Covid-19 vaccine: it is made up of more than 280 ingredients and components that come from suppliers and facilities in 19 different countries. The Oxford-AstraZeneca shot, the "vaccine for the world," as its developers dubbed it, is being made using more than 20 manufacturing sites around the world, including in India, Europe and South Korea. And its ultimate success was built on evidence collected from tens of thousands of volunteers in the United Kingdom, Brazil and South Africa who took part in clinical trials to test its efficacy.

Thankfully too, despite the backlash against South Africa after its scientists shared the vital genetic information on Omicron with the world, researchers around the world are still largely convinced that sharing information as quickly and as widely as possible is by far the best approach. Experts who track such viral data sharing estimate that since the start of the Covid-19 pandemic, more than seven million SARS-CoV-2 genome sequences have been shared via global databases like GISAID by all the 193 countries that are members of the United Nations. For the Omicron variant alone, more than two million genome sequences have been shared by more than 160 countries.

Although it's evidently come too late for Covid-19, governments around the world have also now begun to think about correcting the current dangerous geographical imbalance in vaccine manufacturing capability. Right now, capacity to make vaccines is highly concentrated in just a few countries or regions – notably the United States, Europe, India and China. Each of those has a huge population that will easily absorb billions of doses of vaccines before there are any left over to share. So in an effort to bring vaccine manufacturing home to as many nations as possible, and help create a system in which countries can easily share without risking disadvantaging themselves, the WHO launched an "mRNA hub" technology transfer project in 2021. The plan is that it will help poorer countries manufacture to international standards their own national supplies of high-tech new vaccines at scale. The African Union – an alliance of 55 member states from the continent – has also announced a plan to expand vaccine manufacturing capacity to be able to meet 60 per cent of Africa's needs by 2040.

Scientific collaboration in developing potential defences against or treatments for Covid-19 – across national and regional borders, as well as across disciplines, between institutions and across public and private sectors – was also often remarkable. A study published by researchers in the United States looking at how well scientists had worked together during the first few months of the coronavirus pandemic concluded that work that would normally have taken years was carried out in months "thanks to a coordinated and heroic effort."[27] Part of what enabled this, the study found, was a collective sense of fear felt by scientists about the unknown virus. That fear then propelled them into finding other specialists to team up with so they could set about trying to do something about it. Another part of it was that many scientific research groups already had, to a greater or lesser extent, an international network of specialists that was "ready to be tapped into with a phone call," the researchers found. And part of it was a recognition that no one person or place – be it a company, university, individual scientist or organisation – was able to go it alone, however great their genius or academic calibre. The study also suggested we should "ask ourselves why it had to take such a gigantic human tragedy for us to work together."

It's clear that among the fundamental causes of the terrible inequity of protection the world got during the Covid-19 pandemic – and, for that matter, also during the 2009 H1N1 swine flu pandemic, and during the long battles against measles, pneumonia and other child killers – was that newly-developed protective vaccines, as well as the knowledge, technology and capacity to make them, distribute them and get them into people's arms, were initially both scarce and unevenly spread. Since emerging infectious diseases do not recognise international borders, and the next Disease X will not do so either, to prevent it from becoming the next pandemic, we need to figure out a way of doing things better. But we're not going to do that simply by asking, in vain, for everyone to play fair. We know that's not going to happen. We know human nature is for people to look after number one, and we know national governments will serve their own interests before they turn to the needs of others. So the way to do it is to create the conditions in which people and their governments are more likely to play fair because it's both easier to do so *and* in their interests to do so. When it comes to preventing pandemics, if we properly prepare – by investing in comprehensive global surveillance and alert systems and scientific research and development – then the barriers to sharing are also dramatically reduced. By investing in preparedness, we can make the sharing much less of just the right thing to do and much more of an obvious way to get everyone shielded much faster and to be able to neutralise a Disease X's ability to spiral into a pandemic.

In the early 1970s, the American moral and political philosopher John Rawls was the brains behind a thought experiment, or moral reasoning device, designed to promote impartial decision-making and create fairer systems. Systems where sharing is the natural response because it's the only way for decision makers to also protect their own interests. The device, known as the "original position" and often referred to as the "veil of ignorance," blocks access to the types of information that would feed biases among those creating the system so that they can't tell who in future will benefit most or least from the available options. Rawls, who died in 2002, was awarded an international prize for Logic and Philosophy and, in 1999, the National Humanities Medal in recognition of his work. Presenting the award, former U.S. President Bill Clinton said his thinking had revived and validated the view that "a society in which the most fortunate help the least fortunate is not only a moral society but a logical one."

The Covid-19 pandemic was, in some ways, a vast natural version of Rawls' thought experiment. It put us behind a veil of ignorance about everything from whether we would be more likely to live or die if we were rich, poor, old, young, male, female and so on, to whether the disease would barely reach our countries or would overwhelm them. And, initially, it put us behind a veil of ignorance about which containment measures, medicines or vaccines might prove successful and be worth our governments investing in and introducing. It was when we were all behind that veil – right at the beginning of the novel coronavirus outbreak, before we knew anything that could prompt us to make selfish decisions which would in the end harm everyone and prolong the pandemic – that we would have been most able to design a fair system for all.

Looking ahead, a valuable use of "veil of ignorance" thinking now would be to set about revamping the global system for societies and populations to deal with a looming pandemic. Because it seems to me that we can define ourselves as being in that "original position" right now. We know only a few things: that the next Disease X is out there and that it has the potential to cause a deadly global outbreak. What we don't know – in other words, what is behind the veil – is whether the disease might only affect the Global North or Global South, East or West, rich people or poor people, the old, the very young, men more than women, or exclusively people from one ethnic group or another. We also don't know if it will infect and kill in large numbers, spread silently, rapidly, or slowly, and whether or what kinds of medicines or vaccines might be developed to treat or prevent infection. And we don't know where, when or by whom they might be developed, made and become available. All of which means that now is the best time to create the conditions and build the system that would be most likely to be the fairest for everyone, whatever the next Disease X throws up, and that would be most likely also to neutralise its pandemic threat.

To that end, the World Health Organization is pushing hard for all countries to come together to draw up an International Treaty on Pandemic Prevention, Preparedness and Response, or Pandemic Treaty, that will recognise and embed the value of collective action against common threats. The idea is that the treaty would pre-set rules on sharing knowledge, data, money, technology, know-how and manufacturing capacity, so that when the next Disease X virus emerges and the veil starts to lift as we learn more about it, we have a shared plan of action that can nip its pandemic potential in the bud. After agreeing, against the dramatic backlight of the rapidly spreading Omicron wave of Covid-19, to initiate the Pandemic Treaty process at the November 2021 special session of World Health Assembly, the WHO's 194 member states are aiming to finalise it by May 2024. It's likely to be a tricky endeavour and will demand sustained diplomacy, foresight and leadership. But working together and preparing to share is surely the best way to disarm viral enemies for everyone.

5: PREPARE to Listen

With the Norwegian government being one of the founding members of the Coalition for Epidemic Preparedness Innovations, and with CEPI's headquarters being in Oslo, it's not unusual for Richard to travel to Norway several times a year for government-level meetings, internal or external policy discussions, or global health security conferences. In January 2020, after a somewhat surreal few days trying to get a handle on the newly-emerging coronavirus while mixing with political and business leaders at the World Economic Forum meeting Davos in the Swiss Alps, he had planned a short trip to Norway the following week. On the two-and-a-half-hour long drive down from the Alpine mountain resort to Zurich airport, Richard shared a taxi with Rajeev Venkayya – the fellow Wolverine whom he had met and worked closely with during his time as an adviser at the White House under President George W. Bush. Richard describes Rajeev as one of his closest and cleverest professional and personal friends, and someone he has immense admiration for. "He and I have the kind of unfiltered conversations that you can only have with very good friends for whom you have deep respect," he says.

The two had already in the past couple of weeks been in fairly constant touch via email and instant messaging, with each keen to get a sense of how the other was perceiving the threat posed by the novel coronavirus spreading from Wuhan. Earlier that day, Richard had told Rajeev in an email he already feared this was a "cat-out-of-the-bag" scenario. His "BEST case scenario," he wrote, was that the new virus would be "as lethal and transmissible as SARS" (2003) and that it might be controlled "with more aggressive containment efforts at an earlier point in the epidemic." More likely, however, was that this was a "much more transmissible, cat out of bag, beginning of global pandemic with a relatively lethal pathogen" situation, he wrote. Rajeev, who returns the warm personal admiration for his friend and former colleague, remembers that Swiss Alpine car journey as moment of grim realisation of what was to come. "Richard and I spoke basically the whole time (during that journey)," he told a German documentary maker two years later. It was because he had been able to have that crucial time to talk with Richard at that moment, Rajeev said, that he knew Richard was right: a pandemic was on its way and it was probably already unstoppable. "From the car, I called my wife and told her to order some N95 masks. I said: 'It's coming. It won't be long. It may be six, eight weeks. But it's coming'," Rajeev said.

With this conversation, and that phone call, still running through his mind, and with the rising number of cases of the novel coronavirus heightening his sense of an impending emergency with each passing hour, Richard decided not to cancel the following week's Norway trip. Instead, he would use it to take his Cassandra-style warnings to a multilaterally minded government – and, if possible, to a Head of State who might listen. And do something. On 27th January he met with Dag-Inge Ulstein, an enthusiastic young Christian Democratic politician who was at that time Norway's Minister for International Development. The two met in Oslo's Victoria Terrasse, a 19th Century building complex bedecked with towers, domes and cupolas which was originally designed as a fashionable residential apartment block and was once home to the modernist playwright Henrik Ibsen. The building has dreadful connotations for many. It was taken over and made into a Gestapo and SS headquarters in April 1940 when Nazi Germany occupied Norway during World War II. Many people were imprisoned and tortured there. Some Norwegian prisoners of war were said to have hurled themselves to their deaths from the building's high windows to avoid their Nazi torturers.

Just as Ulstein's meeting with Richard began, Richard's phone rang, stalling the conversation just as it had begun and sending a ripple of tense laughter around the room as Richard explained how his phone was ringing almost constantly these days. With the phone duly silenced, the mood became more sombre. And alarming.

Asked to lay out his assessment of the situation evolving in Wuhan, and his analysis of the potential for global spread of this new respiratory pathogen, Richard didn't hold much back. This was an extremely serious situation, he told minister Ulstein. The novel coronavirus almost certainly had pandemic potential. It had already shown itself to be deadly. There was evidence, too, that it could spread between people asymptomatically – meaning it could not only be a killer, but also an invisible one. There was no vaccine. No treatments. And – if left unchecked – it could be as bad as the Spanish Flu pandemic of 1918. On Norwegian TV news that evening, Ulstein – who, by the way, has a sideline as a lead singer in a Christian pop band called "Elevate" – relayed what he'd heard. "What he (Richard) says is highly alarming," he told viewers. "Lives are at risk on an enormous scale if this develops into the worst case, as we have heard here."

Early the next day, Richard found himself heading to a medieval fortress overlooking a fjord on the outskirts of Oslo. Ulstein clearly also wanted his boss to hear, directly, what Richard had to say. Since the Middle Ages, the Akershus Fortress has served at various times as a city defence post, a royal residence, a military base, a prison, and now as temporary offices for the Norwegian Prime Minister and ministry of defence after the seat of government was blown up by Anders Breivik in a terrorist attack in July 2011.

On 28th January 2020, it was the setting for Richard to meet and talk to Erna Solberg, Norway's then Prime Minister. The two already knew each other slightly, since Erna Solberg was one of the founding mothers of CEPI and had been on stage at its launch in Davos in 2017. Just as he had been the day before, Richard was blunt and explicit in laying out his view of the new and spreading outbreak. Despite the fact that there were only around 12,000 cases and 306 recorded deaths from Covid by 27th January, already by that point, he recalls, his reckoning was that "we had probably lost control of an event that was going to become a gigantic global pandemic." He also believed that, without swift and sweeping social distancing measures, new vaccines or other medical defences to stop it, it could end up killing as many as 50 million people. On his final day at the Davos World Economic Forum meeting – 24th January 2020 – Richard had given an interview to Britain's Channel Four evening news show in which he said more publicly what he had for days been thinking privately. "In 20 years of working on epidemic preparedness, I can't say that I've been more concerned than I am about the current virus," he said. He went on to urge governments and societies to look "right now" at using non-pharmaceutical interventions – those NPIs that he and Carter Mecher had studied in such detail and found to be so effective during the 1918 Spanish Flu. Political leaders and public health advisers should start to consider taking steps such as banning mass gatherings, potentially closing schools, and even introducing travel restrictions to stop the new virus from spreading out of control, he said. He reiterated all of these thoughts and more, with more urgency, to Prime Minister Erna Solberg. "At the time I felt like people thought I had lost it," Richard recalls. "But people's memories may change, of course – because I ended up being right."

Remembering the 28th January meeting at the medieval fortress and relaying her thoughts about it to me two years later, Erna Solberg says Richard's assessment stood way out from the norm. "He gave his estimates on how this could develop, and on the possible deaths. He said that this could have terrible effect in terms of numbers of people dying, and in terms of how many people would get sick. Of course, this was (in January 2020) before we had seen large numbers of casualties and before any people in Europe had even begun getting ill. It was much more alarming than what we had heard from others at that time," she told me. Yet despite Richard's view being more extreme than any others she'd heard, Solberg said, she wished she'd listened to him more – wished she'd heeded his warnings with more gravity and urgency. "It's so important to listen to those who raise the alarm. I'd advise people to do that – especially when we are faced with something we really don't know much about," she said. In other words, when thoughtful, experienced and well-informed people appear to be saying crazy-sounding things, making alarming predictions about an impending disaster – "playing the Cassandra," as Richard put it – it's not at all a bad idea to listen to them.

An official national report into Norway's handling of the Covid-19 pandemic published in April 2022 marked out these meetings as a significant moment for authorities there.[28] Norwegian ministers had in January 2020, the report said, received "professional advice from CEPI," whose chief executive officer Richard Hatchett had told minister Ulstein that the outbreak of the novel SARS-CoV-2 coronavirus in Wuhan was "very likely" to develop into a pandemic. It also noted, with due regard, that Richard had been correct. "Three days later, the World Health Organization declared a global public health crisis," it said.

Richard's second meeting with a Prime Minister that year came in early March in Britain, with Boris Johnson. Three months into the spread of the novel SARS-CoV-2 coronavirus, it had already infected at least 58,500 people around the world and killed almost 2,500 of them. In Italy, a former fellow European Union member state and a popular holiday destination for British tourists in springtime, the WHO's tally of confirmed Covid-19 case numbers had jumped by more than 10,000 in a week – from 4,755 in the week of 2nd March, to 15,274 in the week of 9th March. Italy's death toll too, was exploding.

In Richard's email inbox, the pandemic-watching Wolverines with whom he so regularly discussed unfolding events in what he called "sensemaking" exercises had passed on a terrifying account from an Italian anaesthesiologist who was working at a hospital in the seaport city of Ancona on the Adriatic coast. The Italian doctor had phoned one of his American friends who was an ad hoc additional member on the Wolverines Red Dawn email chain and relayed to him a waking nightmare scenario. Around a fifth of all the patients coming into his hospital in central Italy at that time needed to go straight into the intensive care unit. Doctors and nurses there "are completely overwhelmed and are having to triage 'what to do and who to let die'," the email said. "This is hell... it's a war zone."

Despite his fondness for comparing himself to Britain's wartime Prime Minister, Winston Churchill, Britain's pandemic Prime Minister, Boris Johnson, did not appear to be able in January or February 2020 to recognise the size or significance of the looming infectious disease battle. An investigative report by the BBC's then political editor Laura Kuenssberg quoted one British government source as saying there was a "lack of concern" at the heart of Johnson's administration about the novel coronavirus. He was reported to have said that "the best thing would be to ignore it."[29] So, far from planning for the worst in the first 11 or 12 weeks of 2020, Johnson was still travelling around the country, deliberately messing up his hair, ruffling his tie and skewing his jacket before launching into long and sometimes rambling speeches about Brexit, free trade agreements, and how Britain was ready to lead the world despite exiting from the European Union. He appeared to dismiss calls for travel and trade restrictions and questioned the sanity of global health experts who, like Richard in his Channel Four interview, had suggested schools and even borders may have to close. In a typically grandiose speech delivered at the Royal Naval College in Greenwich on 3rd February – three days after the first two cases of Covid-19 in the United Kingdom were confirmed, and when the WHO was reporting an international toll from the disease of upwards of 23,000 cases and more than 500 deaths – Boris Johnson suggested Britain should resist what he framed as an exaggerated threat of global pandemics. He waxed lyrical about Britain's maritime history and how he was determined to keep the United Kingdom "open for business." This he would fight for, Johnson blustered, even "when there is a risk that new diseases such as coronavirus will trigger a panic and a desire for market segregation that go beyond what is medically rational."

In short, Boris Johnson did not appear to be taking the looming pandemic seriously at all. Having visited a hospital a few days previously where Covid-19 patients were being treated, Johnson blithely told a news conference on 3rd March: "I've been shaking hands continuously. I was at a hospital the other night where I think there were actually a few coronavirus patients, and I shook hands with everybody, you'll be pleased to know."

A few days later, Richard was invited to meet Boris Johnson during a visit to the laboratories of a company called Mologic, located in a technology park in the central English town of Bedford. Mologic is a health technology company which, at that time, was busy developing prototype antibody tests that it hoped would detect whether people had been infected with the novel Covid-19-causing coronavirus. Boris Johnson's government had by that stage already made a big show of buying diagnostics, hoping to show it would be able to get ahead of the game in tracking the outbreak. It bought 3.5 million antibody testing kits from a range of different suppliers before realising it had yet to establish whether or not they worked. As it turned out, Mologic's test did work, and the company went on to be bought out by the multi-billionaires Bill Gates and George Soros in the summer of 2021 as part of a philanthropic project to turn it into a social enterprise.

While Richard arrived uncharacteristically early for the Mologic VIP visit, Boris Johnson was characteristically late. "And when Boris did arrive, he came in and started working the room – like a politician does – glad-handing everybody, shaking hands," Richard recalls. "But when he came up to me, I gave him an elbow – what we used to call the 'Ebola elbow' – rather than offering my hand."

A little taken aback, Johnson asked half-jokingly whether Richard seriously thought the coronavirus situation was really THAT bad. Richard's refusal to switch from an elbow bump to a handshake suggested that indeed he did. Two weeks later, Johnson ordered all pubs, clubs, gyms, theatres and restaurants to close and gave a televised address to the nation to announce a lockdown. A week after that, Johnson's office at Number 10 Downing Street announced that he had tested positive for Covid-19 and was self-isolating. And a week after that, Johnson was admitted to hospital, then put into intensive care and given what he later described as "litres and litres" of oxygen to bolster levels in his blood and lungs. So, yes, it really was THAT bad. Richard and his pandemic-worrying sense-makers were not only "medically rational," they were right.

Clearly, when it comes to predicting, preparing for and preventing pandemics, being right isn't a particularly great position to be in. Making grim forecasts about deadly waves of infectious disease overwhelming countries' health systems and shutting down schools, commerce, travel and economies, and then turning out to be right about them, doesn't make pandemic-watchers feel good. But highlighting these moments in retrospect does help illustrate the importance of encouraging political leaders to listen to experts and to be prepared to accept that their predictions – as scary and even as extreme as they may appear – could well turn out to be right.

It was back in June 2016 when he was campaigning for Brexit that the British politician Michael Gove snapped in an interview that the people of Britain "have had enough of experts from organisations with acronyms saying that they know what is best." Pouring scorn on the experts was a bandwagon Donald Trump had set off on a few months earlier when he was campaigning to become his Republican Party's nominee for the 2016 U.S. presidential elections. "The experts are terrible," Trump declared. "Look at the mess we're in with all these experts that we have." I didn't agree with either Donald Trump or Michael Gove then. The undermining of science and truth by populist politicians has put millions of lives at risk. Expertise in many different forms helped the world get through the coronavirus pandemic better than it would have without it. Listening to experts of all types, in many fields, can prepare us to prevent pandemics in the future. Doctors, nurses, vaccinologists, immunologists, epidemiologists, virologists, pandemic modellers and millions upon millions of other experts with specialist knowledge have all worked on aspects of the global response to Covid-19 and made it better – or less dreadful – than it would have been without them. We can only hope that now that the world has experienced such a devastating event – and one that was to a greater or lesser extent both predicted and mitigated by people with specialist knowledge – that it has become screamingly obvious that experts are worth listening to.

Boris Johnson and his ministers, however, didn't learn to listen to the experts.

After a relative easing off of case and death rates from Covid-19 during the early summer months after Britain's first 11-week lockdown, infections began creeping up again in August and September – not helped by the then Prime Minister and his then Finance Minister Rishi Sunak actively encouraging people to get back to normal, go out to restaurants and to stop working from home and instead return to shops and offices. On 16th September, with daily infection rates topping 4,000, Johnson told a parliamentary committee meeting he did not want a second national lockdown. "I think it would be completely wrong for this country, and we are going to do everything in our power to prevent it," he said.[30]

A few days later, on 21st September, Britain's then Chief Scientific Advisor Patrick Vallance and then Chief Medical Officer Chris Whitty gave a media briefing with a notable absence. Boris Johnson – who had so often appeared between these two experts at joint briefings in the preceding months – was not at a podium this time. This time, the briefing did not feature three men standing side-by-side and a socially-distanced two metres apart at lecterns in Downing Street. This time, it took place in a poorly-lit room with the two advisors only, wearing dark grey suits and sitting grimly behind a large wooden desk. They had bad news. Despite the relative lull in the early summer months, the epidemic was now moving very fast, they said, and Britain was now facing a potential explosion of the epidemic within weeks unless urgent action was taken to halt a rapidly spreading second wave. The UK, at that stage, already had the highest official Covid-19 death toll in Europe and the fifth largest in the world. But things would swiftly get worse still. "If we don't do enough the virus will take off," Whitty said. If things continued as they were, Britain's Covid-19 death toll would "continue to rise, potentially on an exponential curve – that means doubling and doubling and doubling again – and you can quickly move from really quite small numbers to really very large numbers. So we have, in a bad sense, literally turned a corner." Specifically, Vallance and Whitty warned, Britain could see 50,000 new coronavirus cases a day by mid-October, and more than 200 deaths a day by November if no mitigating actions were taken.

It was a briefing that put the two men directly in the firing line. Attacked and pilloried as "Dr Doom and Professor Gloom," or "The Two Horsemen of the Apocalypse," Whitty and Vallance were widely accused of scaremongering. One media report quoted an anonymous Conservative Party member of parliament as saying the two men should be renamed "Witless and Unbalanced."

Sadly, predictably, Boris Johnson, backed by his ministers and many Conservative Party members of parliament, did not introduce any more social restrictions when Whitty and Vallance asked for them. For a further five weeks, he delayed – ignoring numbers that were going very much in the direction his two advisers had predicted. In the end, though, his hand was forced by the pace of the pandemic, and Johnson put the country into a second and then a third national lockdown. The third national closure, on 4th January 2021, came when there were more than 50,000 new cases of infection being reported every day across the United Kingdom.

A report by a thinktank called the Resolution Foundation calculated that if swifter, earlier restrictions had been introduced at the start of autumn – when Whitty and Vallance were under attack for "doom mongering" – and had prevented Covid-19 death rates from rising in December 2020, as many as 27,000 fewer people in England would have died in the epidemic's winter wave. A *Times* newspaper investigation said that more than 1.3 million extra Covid-19 infections were estimated to have spread across the country due to Johnson's refusal to act on his experts' advice.[31] It also cited an analysis by scientists at Imperial College London, who had collated official statistics and the results of mass population sampling, which showed that daily new Covid infection case numbers reached about 45,000 in mid-October 2020, and that the official daily death toll averaged around 430 by mid-November. It turned out, then, that Whitty's and Vallance's predictions were not so unbalanced and apocalyptic after all. On the contrary, they were pretty much right.

During 2020 and 2021, international journalists covering Covid-19 would regularly ask for time to talk to Richard to get his perspective on where the pandemic was heading, how the global vaccine rollout via COVAX was progressing, and what he foresaw as the main potential risks ahead. In one such interview, with a reporter from an

Italian newspaper in September 2021, Richard was asked for his view on Covid-19 variants – mutated versions of the virus that evolve the capability to spread more easily or evade the immune systems defences in new ways. By that stage in the coronavirus pandemic, the globally circulating SARS-CoV-2 virus had already picked up hundreds, probably thousands, of individual mutations and some of those had clustered together to become new variants of the virus. Mutant variants are not specific to SARS-CoV-2, or even to coronaviruses in general. They can occur in any virus, any time it is replicating in an animal or human host. Each time it makes copies of itself, a virus can make a mistake. Many of these mistakes, or mutations, are little more than insignificant one-offs. But some of them – specifically the ones that give the virus a survival advantage – tend to persist. Genetic sequencing information about the emerging mutants posted on global databases meant that the World Health Organization and other disease-monitoring centres could track their evolution and spread, and rank them according to their perceived level of threat.

Initially, when genetic sequences came in showing significant changes all appearing in a cluster of cases of infection, variants were designated a Variant of Interest, or VOI. Then, if more cases began to emerge, in more locations, or if the mutations were found to be in crucial parts of the virus that could make it more transmissible or more dangerous, the designation would be taken up a rank to Variant of Concern, or VOC. By September 2021, five new Variants of Concern had already been detected and identified as worrying versions of SARS-CoV-2. The first – later named Alpha under a new, Greek-alphabet-based variant naming system designed by the WHO to be simple, non-stigmatising, and easy to say and remember – was detected in south-east England in September 2020. Others, duly named Beta, Gamma, Delta and then Lambda, were found not long after that in South Africa, Brazil, India and Peru.

The Delta variant – which was around 50 per cent more transmissible (contagious) than Alpha, which itself was around 50 per cent more contagious than the original SARS-CoV-2 virus – swiftly became the dominant form of the virus worldwide. Starting in India, it swept through the rest of Asia and spread to Europe and the U.S. within months. Having been officially recognised and named by the WHO at the end of May 2021, it had spread to almost 180 countries by end of November the same year.

Despite the clear and regular emergence of these new mutant SARS-CoV-2 variants, however, there was a sense, each time, that people wanted to revert to the relatively comforting view that this was the worst the novel coronavirus could be. After obliterating its predecessor variants, Delta was dubbed the fastest, fittest and most formidable version of the coronavirus the world had encountered. And that was correct. At that point.

Cassandra-minded pandemic preparedness experts like Richard, however, were very reluctant to be reassured that Delta was the worst SARS-CoV-2 could come up with. They were watching the virus circulating in vast numbers of people – millions upon millions across the world at any one time – and from time to time encountering immune pressure of one sort of another, either in a person who was partially or fully vaccinated, or in one who had some immunity from a previous Covid infection. Capturing this scene and recognising its threat, in the interview with the Italian journalist in September 2021, Richard gave a warning: the way the world was vaccinating its people – or rather vaccinating only some of them – against Covid-19, he said, was creating a perfect environment for more variants to evolve, and to be more threatening. "If we wanted to design a world that would increase the probability of emerging variants, it would be to do this: we'd have some very highly vaccinated and immune populations in some parts of the world, and we'd have huge swathes of the world where very few if any people are getting vaccines, so there is lots of transmission," he said. "Right now, we've got both of those things. We're in a lot of risk." Barely two months later, the world was hit by the Omicron variant.

The earliest confirmed cases of the Omicron variant of Covid-19, which had more than 60 mutations, including 42 changes in the infamous spike protein it uses to break in to and infect human cells, were detected in Botswana and South Africa in the first half of November 2021. That was also when the new mutant hit world headlines, although scientists now think it is likely to have begun spreading in people a month before that, in early October. It is also now judged to be one of the most – if not the most – contagious human respiratory viruses ever known. Omicron had fulfilled, in an intensely depressing and precise way, the predictions of experts like Richard who had repeatedly cautioned that allowing the virus to spread in some countries and regions, by failing to get them access to vaccines, would super-charge its ability to evolve and mutate. Having only had access to enough vaccines to immunise less than a quarter of their populations, Botswana and South Africa provided a perfect and highly fertile breeding ground for Omicron. Again, predicting what appeared to be the worst-case scenario turned out to be predicting reality.

Just for the record, it's also worth noting that at the Global Pandemic Preparedness Summit co-hosted by CEPI and the UK government in March 2022, Chris Whitty's prediction was that, even with the Covid-19 pandemic barely behind us, the risk of future of significant disease outbreaks is "relatively high," as is the threat of pandemics.

Largely because of Omicron, the intensely contagious variant of Covid-19 which emerged in late 2021 and caused a fresh tsunami of infections around the world over Christmas into the New Year, the annual Davos gathering of global leaders at the World Economic Forum was postponed from January to May 2022. This meant that when world leaders got together in the Swiss Alpine town for the first time since January 2020, there was another curious disease threat occupying their minds. "This is the first time that we have gathered again in Davos since the 2020 meeting – and we find ourselves facing another dangerous disease threat," Richard said in an interview with CNBC at the time. "We should understand what that signifies, which is the world is beginning to move around again and infectious disease is beginning to move around with us." Unlike the novel and mysterious coronavirus SARS-CoV-2, however, this virus was not new. Rare and curious, yes, but for a start, it already had a name: monkeypox.

While few people outside of West and Central Africa had ever heard of it until the spring of 2022, monkeypox is a known human pathogen. It causes small outbreaks regularly in central and west African countries such as the Democratic Republic of Congo, Cameroon and Nigeria, and was only extremely rarely seen in Europe, the Americas or Australasia.

The first known human case of monkeypox was detected in 1970 in a child in the Democratic Republic of Congo. The rash of blisters on the boy's body initially struck terror into the hearts of doctors, who feared it might be a reappearance in Congo of the by then eliminated and much more deadly smallpox virus. Further testing showed this infection was not smallpox, but monkeypox – a far less dangerous virus that had first been identified in monkeys 12 years earlier. Both monkeypox and smallpox belong to a family of viruses known as orthopoxviruses – a family that also includes viruses that cause cowpox, horsepox, racoonpox, camelpox and several others. Some of these viruses – although not all – have shown the ability to jump from animals and sporadically infect people. Monkeypox is much less contagious than smallpox and also causes significantly less severe illness. While smallpox, before it was eradicated, had a death rate of up to 30 per cent, deaths from monkeypox infection are uncommon. In the first half of 2022, for example, there were around 70 deaths from monkeypox across five African countries, according to the World Health Organization.

The orthopoxvirus family, however, is one which epidemiologists and pandemic preparedness experts have been keeping an eye on, and warning about, about for several years. In a scientific paper published in September 2020, a year and a half before the World Health Organization began in May 2022 to report scores of "atypical" cases of monkeypox infection cropping up in unusual locations around world, epidemiologists from the Institut Pasteur in France had published a paper warning about it. Because the geographic spread of monkeypox cases had already begun expanding beyond the forests of central Africa, they wrote, "the epidemic potential of monkeypox will continue increasing." It looks like they, too, were right.

The 2022 international monkeypox outbreak is also a case in point proving several other uncomfortable truths: that deforestation and intensive agriculture are creating ever more opportunities for the types of animal-to-human contacts that allow new zoonotic diseases to emerge; that rapid urbanisation and globalisation increase the likelihood and speed of human-to-human disease transmission and make epidemic and pandemic threats more frequent; and that viruses do not respect borders of any kind.

The first case of monkeypox in the 2022 outbreak was reported in Britain on 7th May. Scores more cases were detected across Europe, the United States and Canada in the subsequent few weeks. By the end of that month, the global case count was in the hundreds, but there were no monkeypox deaths in the outbreak. Given the world's heightened awareness of infectious diseases at the time these first atypical cases were spreading in non-endemic countries, it was not surprising that alarms started to sound. Within days media headlines began asking "Is monkeypox the new Covid-19?" and "Will monkeypox become a pandemic?" In the United States, where around 10 confirmed cases were reported in May 2022, President Joe Biden said monkeypox was now "something everyone should be concerned about."

Concerned, yes. But not pandemic-scale alarmed. Because as it happens, with monkeypox, we've already done the right thing. For monkeypox in 2022 was a perfect example of how we can contain disease epidemics and prevent them from spawning global pandemics *by being properly prepared.* It's events like these that make it crystal clear that we are absolutely right to be cajoling political leaders, and their finance and defence ministers, to prepare for the unknown to get ahead emerging diseases rather than trying to chase them down. Largely thanks to Edward Jenner's work on developing the world's first and highly effective smallpox vaccine, and thanks in part to some governments recognising the potential of smallpox as a bioterrorism threat, the world's scientists and pharmaceutical companies have developed and created major defences against smallpox. And because smallpox and monkeypox both belong to the same viral family – in the same way as MERS, SARS and Covid-19 do in the coronavirus family – the tools and defences we have developed against smallpox can be deployed to contain outbreaks of its far less lethal cousin, monkeypox. There are at least two smallpox vaccines, fully developed, licensed, and in some places stockpiled in relatively large volumes. These smallpox vaccines can be as much as 85 per cent effective in blocking infection with the monkeypox virus if they are given before people are exposed to it. And there are potential treatments on the shelf too. Although most cases of monkeypox infection are mild and generally clear up on their own after a few weeks without the need for treatments, antiviral medicines developed for use in treating smallpox infections – and again stockpiled in case they are needed in an emergency – can potentially be used to treat monkeypox patients who get more seriously sick.

All of that means that, when it comes monkeypox and any other known orthopoxviruses that crop up or reappear in future, the world is in a much stronger position to be able to respond to and contain outbreaks than it was to be able to contain Ebola, for example, or Covid-19. More than anything, that's because we had done the scientific discovery and development work – and put in all the investment and innovation effort required – ahead of time. The way it works is like this: the smallpox virus acts as a prototype for the orthopox viral family. This means that through a process known as cross-protection, the vaccines against smallpox can also protect against all the other known orthopoxviruses in that family. The monkeypox outbreak of 2022 was the poster child for this strategy, and showed the enormous value in adopting the same approach for all of the 25 or so viral families known to be able to infect people. That way, we build a library of pre-prepared vaccines against at least one or two prototype viruses from each of these families. If, like the smallpox vaccines of today, these pre-prepared vaccines are fully tested for safety, and have already been shown to be able to protect against the prototype viruses, they can be taken off the shelf, and adapted if necessary, to be used to protect people from a re-emerging, newly emerging, or totally novel virus in that same family.

"With monkeypox, we had the tools that we need just as the epidemic started," said Richard. "Here is a disease that the world has not focused a lot of attention on, but one that we are nevertheless ready for because we invested in developing countermeasures against the prototype orthopoxvirus, which is smallpox." And because of that, we can be confident that we're not going to have a monkeypox pandemic. "This is not a disease that is going to sweep around the world and have people cowering in their homes like Covid-19 did."

6: PREPARE to Fail

On 3rd February 2020, just as the SARS-CoV-2 virus that causes Covid-19 began its rapid, long and lethal global march, the United States National Institute of Allergy and Infectious Diseases – the scientific research institution led by the now world-famous Dr Anthony Fauci – issued a depressing but relatively little-noticed announcement.[32] An expensive and long-running clinical trial involving more than 5,400 volunteers in South Africa to test the potential of a candidate vaccine against HIV, the human immunodeficiency virus that causes AIDS, had proved useless. It was time to call a halt to the $104 million study, the NIAID statement said. The trial, code-named HVTN 702 or Uhambo, had gotten peoples' hopes up. It had been built on some already promising evidence from another curiously-named trial – RV144 – that had been conducted among thousands of people in Thailand between 2003 and 2009 and had been the only trial thus far to appear to show that an HIV vaccine could at least create a partially protective immune response.

The vaccine being tested in the Uhambo trial was based on a live bird virus called canarypox – a virus that scientists have discovered can infect human cells without causing any disease. That ability made it a promising so-called vector virus for a vaccine – one that carries material into human cells that then induces them to make several key proteins of the HIV virus. The hope was that those proteins would in turn stimulate the immune system to recognise and defend the body against any future HIV infection. That hope faded fast, however. Four years in to the trial, the study's independent monitoring board unsealed some of the data collected from the 14 sites participating and found that the experimental shot was making no difference. The trial had been due to run for six years, until July 2022, but by early 2020, the board concluded that continuing it would be futile. "There's absolutely no evidence of efficacy" was the blunt assessment made by Glenda Gray, an AIDS doctor, scientist and activist who co-led the study and is also president of the South African Medical Research Council. The number of people newly infected with HIV in the trial's placebo control group was almost the same as the number in the group that had been given the experimental canarypox-vectored vaccine. The candidate vaccine had failed. And for AIDS vaccine researchers across the world, it was back to the drawing board. Again.

Again – because unfortunately the halting of the Uhambo trial was just the latest in a sorrowful string of failures that has characterised the search for a vaccine to prevent HIV. More than a decade earlier, in 2008, two separate trials of an experimental AIDS vaccine designed by the American pharmaceutical company Merck & Co and called MRK-Ad5 were also stopped after a mid-stage data analysis showed very worrying results. Merck is one of the most successful vaccine developers in the world and is behind eight of the 14 now routine vaccines currently recommended for children in the United States. But these AIDS vaccine trials – one called STEP which involved more than a thousand men and women participating in Australia, the Caribbean, and North and South America, and another called Phambili which involved another thousand or so volunteers in South Africa – found not only that the candidate vaccine gave no protection against HIV, but that it might even have made the risk of infection higher. More people in the group vaccinated with the experimental shot went on to become infected with HIV than did in the control group that got a placebo vaccine. The candidate vaccine had completely failed. Again.

Yet again – because the reality of the situation is, that despite the fact that the first trial of an HIV vaccine began some 35 years ago in 1987, the world's scientists and pharmaceutical companies have consistently failed to come up with one that works. This list of failure after failure is a long way from the now ludicrous-seeming prediction in the 1980s by the then U.S. Secretary of State for Health and Human Services, Margaret Heckler. In April 1984, just three years after the HIV epidemic began, she gave a news conference at which she said scientists had already started work on developing an AIDS vaccine and hoped "to have such a vaccine ready for testing in approximately two years."[33] (It's worth noting that Dr Anthony Fauci, who back then was just starting out on his career in fighting infectious disease epidemics, was distinctly reticent about making any kind of positive prediction for the discovery and development of an AIDS vaccine. He told reporters at a medical convention in Atlanta at the time: "To be perfectly honest, we don't have any idea how long it's going to take to develop a vaccine, if indeed we will be able to develop a vaccine.")

Yet the failure-ridden search for an AIDS vaccine has delivered far more than just a decades-long series of trips down blind alleys ending in frustrating dead ends. Granted, there have been multiple false starts, and many dashed hopes. But there has also been huge progress. The search has brought scientists a wealth of new knowledge about how and why vaccines work or don't work, and about how viruses like HIV – which, like the one that causes Covid-19, is an RNA virus – can evade immune defences, mutate, and find news ways of getting around both the body's innate and its adaptive immune system. It has taught vaccine developers about the up- and downsides of a range of viral vectors, including the Adenovirus 5 vector that was used by the Russian developers of the Sputnik V Covid-19 vaccine, and the Adenovirus 26 vector which is part of the successful Covid-19 shot developed by Johnson & Johnson. All this work – littered as it is with failures – has became a critical part of the foundations for the development of vaccines against Covid-19. In essence, we would not be where we are today without the long trail of failures in HIV vaccine research. The work has achieved much of what Richard describes as "solving the problems of vaccinology" in advance, paving the way for speedier, more successful science to come. "If you understand the vaccinology, you're putting yourself in a position to have a much higher degree of success than if you're starting from scratch," he says. "With HIV, we spent 30 years or more testing out different vaccine concepts and that has, in part, helped lead us to a situation where we have fewer failures with Covid vaccines than we would have had."

The proof is in the numbers. As of the middle of February 2022, the World Health Organization's Covid-19 Candidate Vaccine Landscape and Tracker was listing 143 candidate vaccines undergoing human clinical trials or already in use, and 195 more experimental vaccines at the preclinical development stage.[34] According to an analysis published in the journal *Health Policy and Technology* in March 2022 exploring how HIV vaccine failures impacted Covid vaccine research, among the 143 candidate Covid vaccines in the clinical phase, 119, or 83 per cent, involved technologies that could be traced back to prototypes tested in HIV vaccine trials.[35] The analysis concluded not just that HIV and AIDS research had added to fundamental scientific knowledge about viruses, vaccines and antibodies, but specifically that "the repeated failures of HIV vaccine clinical trials have served as a critical stimulus to the development of successful vaccine technologies today."

To successfully pursue the science we need to build a library of potential vaccine defences against future Disease Xs with pandemic potential, we must expect to have to accept many failures along the way. And to succeed, we need to be prepared to learn from them. All of which is why making friends with episodic failure is an essential pursuit for Melanie Saville, CEPI's Executive Director of Vaccine Research and Development and the woman in charge of its vast portfolio of vaccine development projects against known and unknown diseases. Melanie is a little embarrassed at others' perceptions of her work ethic as remarkable. She puts her drive and determination as an adult down, in part, to having been seen as an underachiever as a child at school. Born and brought up on the Channel Island of Jersey, she left school at 16 after her teachers, she says, pigeonholed her into the "not particularly academically gifted" category. Initially accepting this verdict on her intellectual abilities, she headed to a somewhat old-fashioned and prim establishment in the southern English county of Somerset: Norland College – a training school college for young women who want to go into service as live-in childcarers for wealthy families. Graduating from there, Melanie duly became a Norland Nanny, complete with a uniform of a beige belted dress, ankle length brown trenchcoat and a brown felt hat embroidered with a gold "N" (for Norland, or nanny, or both). Norland College is renowned worldwide for producing nannies for the rich and famous. Often referred to as "the real life Mary Poppins," Norland graduates have looked after everyone from Britain's second in line to the throne, Prince George of Cambridge, and his younger siblings Princess Charlotte and Prince Louis, to the offspring of the rock legend Mick Jagger.

Melanie didn't get to hang out with royalty or rock stars, but, working as a live-in nanny to a young baby girl in a well-to-do family in London, she nevertheless learned a lot about life, about how to organise and work with other people, and about herself. Most strikingly, she quickly realised she wanted more from a career than the kind of proxy motherhood the nanny position offered. So she decided to go back to school – a night school where she could study after her day's work for the A levels she needed to be able to go and train as a doctor. At that stage, she thought she wanted to specialise in treating children with cancer. But when, having worked relentlessly to get herself a place to read medicine at University College London, she took a course in molecular biology in her first month of studies, she fell in love. "When I did that first molecular biology course, it got me totally hooked on viruses," she recalled.

From there, Melanie went on to work as a virologist and in vaccine development for a range of nasty diseases, including rabies, Japanese Encephalitis, Dengue and Ebola and multiple types of flu. Her first pharmaceutical industry position was at the drug company Wyeth – a vaccine maker subsequently taken over by Pfizer. She moved to the French drugmaker Sanofi Pasteur in 2004 for 11 years, and then to Janssen Pharmaceuticals, the Belgian drug company owned by America's

Johnson & Johnson. At CEPI, where Melanie joined Richard as part of the leadership team in 2017, she is known for being super bright, taking few vacation days and being "interested in absolutely everything." She's also very much an advocate of "Let's fail, and let's fail fast." Melanie herself adds to that: "And let's fail often."

No surprise then, that when the first cases of the novel coronavirus emerged in China in January 2020, and Melanie was approached by Richard in CEPI's London offices with a New Year's greeting that went something like: "Melanie, we have a problem," Melanie was all ears.

That same day, she joined a meeting with the World Health Organization and other global health experts to hear more about what she then called "this mysterious virus in China." "At the time we weren't even sure what it was, but it was obviously an outbreak that the international community needed to act on," Melanie says. "We knew we needed to act quickly."

Starting with those super-fast investments in experimental Covid-19 vaccine development projects at Inovio, Moderna and the University of Queensland, Melanie and her team went on over the following two years to expand the number of CEPI-funded potential coronavirus vaccines to a total of 14 – making CEPI's Covid vaccine development portfolio one of the world's largest and most diverse in terms of geography and technology.

The process involved reviewing at first scores, and ultimately more than a hundred applications for funding by start-ups, research groups, biotechs and others who came to CEPI with their ideas for how to make an effective Covid-19 vaccine and sought the coalition's backing. In an email exchange with one potential collaborative partner on 22nd February 2020, Richard stepped in to the conversation, noting that he was keen to give Melanie "a bit of a breather" after she and her team had reviewed a total of 48 applications from potential coronavirus vaccine developers from all over the world in less than a week.

Mel's team's job was to pick out those with highest likelihood of success, but also to ensure the world's pandemic defence eggs were not all thrown into one basket. The mRNA technology that Moderna was planning to use in its candidate vaccine had never yet proven successful, and another applicant, CureVac, was planning to use a similar technology but with a slightly different form of mRNA. The University of Queensland project, the one that used a "molecular clamp," was also an innovative – and hence risky – approach. And Inovio, which got a grant of up to $22.5 million from CEPI for its Covid-19 work, was also using a relatively new type of DNA technology in its proposed vaccine candidate.

Then there were the viral vector techniques being proposed by other research groups like the team at Oxford University led by Dame Sarah Gilbert and Professor Adrian Hill. These viral vector techniques also had a history of mixed results – with some failures in HIV and a notable success in the (eventual) development of an Ebola vaccine. One vector-based project came from an Austrian biotech company called Themis, which along with France's Institut Pasteur had developed an experimental vaccine containing a weakened and safe version of the measles virus to serve as a carrier for genes from the SARS-CoV-2 virus. In March 2020, when CEPI had already provided initial funding to seven other projects, Melanie and her team also picked this Themis project as one of CEPI's "multiple shots on goal."

In vaccines – particularly ones being developed at speed during an evolving global pandemic – there are many ways a project can fail. One of the key reasons for the world needing to start with such a huge number of Covid-19 vaccine development projects was that failures can come from just about anywhere along the way. Sometimes it's the science. Sometimes it could be a technical issue such as a potential cold-chain storage problem. Other times it could be a human partnership-type problem, where a lack of trust or abundance of suspicion on the part of one or other project partner leads to the end of what might otherwise have been a successful relationship. There are manufacturing failures, policy-linked failures and financing failures too. "You constantly have to question yourself and take into account what's happening in the world, what's changing. Is this still relevant for the epidemic as it is now? Has the outbreak moved on or become different? Is there a newer and better technology out there now that wasn't there at the beginning? All of that has to be integrated into the dynamic of building a potential vaccine portfolio," Melanie says. "That's also why failing fast is so important. You don't want to drag things out, and you don't want to waste public funds and effort defending a project that shouldn't be supported any more."

Of course, it's not easy to argue that failure is an entirely positive thing. In many instances it's very much the opposite. It can take up time, waste money, damage reputations, make researchers or others working on a project feel dejected and inclined to give up. In the worst instances, it can be dangerous and put lives at risk. But for a threatened world preparing to be able to stop the next pandemic by developing safe and effective new vaccines within 100 days, multiple failures along the way are inevitable. And if failure is managed well, it can be very constructive.

Melanie knew there was a good chance that many – possibly most – of the 14 CEPI-backed Covid-19 vaccine development projects were likely to fail, or not to progress as quickly or successfully as might have been hoped. And many of them did. Themis among them. By early 2022, of the 14 Covid vaccine development projects CEPI had begun, six were still ongoing, three had successfully made it through to full development with licensed vaccines on the market, and five had failed. That's a scorecard Mel is very comfortable with. Failure doesn't worry her. "Not at all," she says. "We should always have failures. We're not doing our job properly if we don't."

A common trait of some of the world's greatest innovators, thinkers and entrepreneurs is a mindset that understands risk and embraces failure. People like Bill Gates, the co-founder of Microsoft, a multi-billionaire, and now one of the largest funders of global health and development programmes around the world, have spent their lives failing far more often than they succeed. But they refuse to be put off by setbacks. In one of his first books, *Business @ the Speed of Thought*, first published in 2009, Gates says we should embrace failure to be able to learn from it.[36] "Once you embrace unpleasant news not as negative but as evidence of a need for change, you aren't defeated by it. You're learning from it. It's all in how you approach failures. And believe me, we know a lot about failures at Microsoft."

It wasn't just at Microsoft that Gates made friends with failure. Since Gates and his ex-wife Melinda French Gates launched the vast Bill & Melinda Gates Foundation in 2000, the philanthropist has relentlessly pushed on multiple fronts to try to solve the world's most stubborn and painful problems. The Foundation has grown to be one of the most influential philanthropic funds in the world, and has an endowment of about $53 billion. Over my years as a global health correspondent, I interviewed Bill Gates many times, often in person – although our last interview was over Zoom as it happened to also coincide with a day in January 2021 when I first came down with Covid-19. Gates never fails to set his sights high. And he rarely misses an opportunity to emphasise that, despite working in one of the most distressing fields of global health, with babies and toddlers dying of malaria, polio, pneumonia and diarrhoea, he is an optimist.

Over the years, Gates has publicly pledged two major global health missions. The first is to eradicate malaria – something which has so far proved impossible despite having scores of ways to prevent infection, score mores ways to treat it, and now, thanks in large part to funding from the Gates Foundation and the persistence of Joe Cohen and his team, the world's first malaria vaccine. The second is to wipe out polio – a viral disease that in 1988 was endemic in 125 countries and caused paralysis in almost 350,000 people, mostly children, every year, but is now only causing a few dozen cases a year worldwide.

In 2012, noting that there had been a 99 per cent reduction in cases of polio worldwide since a global eradication campaign was launched in 1988, Gates predicted the world would rid itself of polio completely by 2018. It got close, but didn't quite make it. In 2018, there were 33 cases of polio recorded across the world, and only six in the first three months of 2019. This led Gates, with renewed determination and optimism, to forecast that the polio endgame was near. Years of conflict in Afghanistan, however, have repeatedly interrupted polio immunisation campaigns there and in neighbouring Pakistan. And in 2020 and 2021, when the Covid-19 pandemic hit and when NATO withdrew from Afghanistan in the face of Taliban advances, polio case numbers rose back up over a hundred in Afghanistan and Pakistan.

Malaria is at least as difficult a nut to crack, if not more so. Bill and Melinda Gates first called for the world to commit to eradicating malaria – that is, wiping it out completely in every country – in 2007. Back then, reviving use of the "e-word" in relation to malaria was seen by many in global health circles as naive, and even dangerous, after a previous project, the Global Malaria Eradication Programme, had been abandoned due to failure in 1969. Undeterred, Bill and Melinda Gates declared they "would not stop working until malaria is eradicated." Aspiring to anything less would be "timid," they said. Then, in a blog in 2014, Gates put a timescale on the ambition. He said he believed the world was in a position to be able to banish all malaria within a generation. "This is one of the greatest opportunities the global health world has ever had," he said, adding that he was so optimistic about it that his Foundation had decided to increase its malaria budget by 30 per cent.

Whether or not Gates will fail to hit his "within a generation" malaria eradication target remains to be seen. Equally, the endgame for polio eradication could be a painfully long and slow one. But one thing is almost certain: Gates, his Foundation and the many other malaria activists and end-polio campaigners he works with and helps to fund will not be put off by failures. They will seek out scientists to work on new and better types of malaria vaccines. And they will keep working on figuring out ways to get polio vaccines to every child in every city, town or village anywhere that the virus pops up. They will cajole other philanthropic foundations and wealthy governments to keep up the fight, and funding, too. In doing so, while the campaigns may not achieve full success or meet their deadlines, they will, step by step, mean fewer and fewer babies and children are left at risk of being killed by a mosquito-borne disease before they reach their fifth birthdays. And they will mean that each year, more and more children who might otherwise have been paralysed by polio will instead be able to walk and run and live active lives. To me, these are among the most laudable cases of shooting for the moon, missing, but landing among the stars. Such projects are fraught with failure, but deliver untold success.

It's something of a twist of fate that, in the two years they have spent collaborating more closely and more often than ever before, Richard and

Seth got very little face-to-face time together. After their surreal time in Davos, Switzerland, at the very start of 2020, when they each became more aware of the looming pandemic as the hours ticked by, they were barely able to meet again in person until around two years later, when the travel restrictions enforced by many countries began to be lifted. Richard says the two became very close – "almost like blood brothers" – during the intense work of the Covid-19 pandemic response and the setting up and co-leading of COVAX, the global Covid vaccine sharing mechanism established to try and ensure equitable and timely distribution of the newly developed vaccines. "We've been in the trenches together, so to speak, so we are able to be honest with each other. We trust each other to say things that we aren't necessarily willing to say more broadly or more publicly," Richard said about the relationship.

The two did meet in early 2022, at the CEPI-hosted Global Pandemic Preparedness Summit in London in March, and went on to have several days of in-person meetings with other the co-leaders of COVAX at the Wellcome Trust building on London's Euston Road. This second gathering was what the COVAX co-leaders called a "retreat," and it was the first time the whole of COVAX'S leadership team had ever come together in person. As a kind of icebreaker for the first morning of the two-and-a-half-day meeting, Julika Erfurt, the event's coordinator, laid out a spread of craft materials suitable for a group of pre-schoolers. She then directed each member of the team to draw a name out of a hat, and handed each of them a small circular piece of white card and a length of ribbon. With these, she said, and with any of the colourful textiles and craft materials made available, they should design and make a paper medal for the person whose name they had drawn from the hat. Immediately inspired with an idea for his recipient, and filled with enthusiasm for the artistic endeavour, Richard chose felt tip pens in pink, orange, blue, green, purple and yellow; a piece of bright green felt; one black and white plastic stick-on goggle eye; some strips of glittered paper; and a little glue. From these he fashioned what was meant to resemble a small green donkey, but came out looking more like a mutant, one-eyed, long-legged frog with hefty horns and a large beak. Creative art is not one of Richard's strengths. But undeterred, he added shiny blue and green stripes to the creature's torso and dotted sparkly gold and silver stars around it. Using the coloured pens in a letter-by-letter rainbow rotation, he wrote across the top of the cardboard medal: "Piñata In Chief."

Richard's medal was for Seth. It was designed as a piñata – a colourfully decorated animal-shaped cardboard container full of candy that blindfolded guests at a party take turns to beat with a stick. It was in recognition of and in empathy with the seemingly relentless attacks from observers in the media, governments and from non-governmental organisations which Seth had been enduring over the past few years as the public face of COVAX. "I think he has felt abused and maltreated," Richard says. "And I empathised with him to a great extent in that." Richard had stolen the sympathetically-intentioned piñata label from a former colleague from his White House days, Lisa Koonin. Koonin, who had worked at the U.S. Centers for Disease Control and Prevention and collaborated with Richard and others on the 2006/2007 study re-examining the 1918 Spanish Flu, had been the first to coin the nickname "piñata" – for Richard. This, he says, was "because I used to have to stand up in front of experts at the CDC and other organisations and defend what seemed to them like crazy ideas." Those crazy ideas, he explains, were the notions that non-pharmaceutical interventions such as social distancing, work-from-home orders and school closures could and should be put in place to slow the spread of infectious disease outbreaks – especially respiratory ones like flu. "The people listening to my presentations would just whack away at me with the limited number of sticks they could think of on the spot, and then with as many more as they could think of later."

Seth's beatings came mostly in 2020 and 2021. And they were mostly to do with how and why COVAX, the global vaccine sharing facility he and Richard had first sketched out over whiskey and nachos in a bar in the Swiss Alps, was struggling to meet its ambitious targets.

In April of 2021, people in countries across Africa were eagerly awaiting deliveries of the new Covid-19 vaccines they had watched being rolled out with such urgency and enthusiasm across Britain, Europe and the United States. Frustrated at being several months behind in getting doses, they were nevertheless also encouraged by the initial trickle of deliveries that had begun to arrive: Ghana and Cote D'Ivoire, in West Africa, got just over half a million Covid-19 vaccines each in deliveries in mid-February; Rwanda, in East-Central Africa, got around 350,000 doses in a first shipment in early March; and, on what GAVI's country manager for South Sudan called a "historic day" – 25th March 2021 – 132,000 doses of the AstraZeneca Covid-19 vaccine arrived at Juba International Airport for the start of the rollout there.

But the celebratory scenes and pictures of wooden delivery pallets piled high with boxes of vaccines being unloaded at airports were rapidly to become less familiar images. By the beginning of April 2021, the COVAX scheme was having to tell expectant countries that their next deliveries, and sometimes their first ones, weren't coming yet. The problem was one of supply. And the supplies were almost all coming from India. That was until they stopped, abruptly.

On 25th March 2021, India, a country of 1.4 billion people, reported that 53,000 people had been newly infected with Covid-19 in the preceding 24-hour period. This was the highest number since October 2020, when the first wave of the novel coronavirus pandemic had swept through this huge, populous country. The mutant Delta variant of Covid-19 was doing its worst – spreading like wildfire and killing tens of thousands of people in a matter of weeks. Hospitals were overwhelmed with patients, and then overrun with desperate relatives begging for oxygen to help their loved ones breathe their last breaths. With doctors and nurses at breaking point – either due to overwork, being sick themselves, or having to care for infected family members – hospital wards had fewer staff than patients. And at times, more of those patients were dead than alive. In Delhi, India's vast capital city, crematoriums ran out of firewood. A later United Nations report was to find that at least 240,000 people died of Covid-19 in India in between April and June 2021.[37]

That 25th March date was also, as it turned out, the same day on which India's government decided that as long as India's own population was still mostly unprotected, exports of vaccines to protect other countries had to stop. Up until that point, only the elderly and people aged over

45 who had other health conditions had been eligible for Covid vaccination in India. So the 25th March de facto export ban was "about survival," according to Kiran Mazumdar-Shaw, chief executive of an Indian biotech company called Biocon, who was interviewed at the time by the *Financial Times*. "India has been charitable in exporting 60 million doses thus far. It's now time to vaccinate Indians," she said.

It was an understandable move, and arguably the only one India's government could make to try to better protect its people at that terrible time. But for COVAX and the scores of poor countries relying on its supplies of Covid vaccines, it was a disaster. Looking back at the situation from the point of view of its implications for COVAX, for its dependents, and ultimately for the slower sharing of vaccines around the world and all the knock-on effects of that inequity, Richard identifies the reliance on the Serum Institute of India to provide the vast majority of Covid-19 vaccines to COVAX as a "giant mistake." "We were wholly dependent on a single point of failure for most of the countries that were relying the most on COVAX to get Covid vaccines," he says. "And then that single point of failure failed."

What led COVAX to that point, though, was a combination of factors that, even with hindsight, seem unavoidable. One central factor was money. Because COVAX was conceived and created on the fly, in the midst of a vast and rapidly worsening infectious disease pandemic that was already closing down economies and battering stock markets, it didn't have any financial resources up front. To be able to secure the means to buy doses of future Covid vaccines in large enough quantities to deliver on its promise to help protect the most vulnerable populations across the whole world, COVAX needed billions of dollars, fast. Recognising this urgent need, the organisation's leaders approached various global financial institutions seeking emergency funds to put money at risk to make and buy Covid vaccines that were still in development and might at any point fail. Needless to say, those institutions were not structured to be able to release money on those kinds of terms. Raising the money was hard and time-consuming.

Then, when it came to spending that money as wisely and as broadly as was possible at a time when still none of the experimental Covid vaccines in development had yet proven successful, COVAX's leaders had to decide how to place their bets. Here, Richard says with hindsight, was where taking greater risk would have been a smarter approach. "Because of resource constraints, we wanted every dollar that we had to go as far as it could go," he said. While the data rapidly coming out from early and mid-stage trials of the new candidate mRNA vaccines being developed by Pfizer-BioNTech and by Moderna were looking promising, those shots were also looking as though they would be both expensive to produce and difficult to deliver and distribute in poorer countries in the global South where cold-chain capabilities and health system resources are limited. "So we put all of our bets on the cheaper vaccine from AstraZeneca made at the Serum Institute," says Richard. "But we hadn't sufficiently hedged our bets. So in doing that, we put ourselves at risk of a single point of failure."

In referring to this misjudgement as "a giant mistake," Richard reflects back on a caution he often heard from a former colleague during his second stint working on pandemic preparedness at the White House in 2009. Heidi Avery, who was at that time Deputy Advisor on Homeland Security to President Barack Obama and was known to her colleagues as a true hard-ass boss, had observed and articulated on many occasions the following: that, through the actions they take to avoid something they fear, people or organisations often unwittingly create the very thing they are most afraid of. On reflection, Richard is pretty sure this was the behavioural and decision-making trap he and other COVAX leaders fell into when they – as he puts it – "dithered" about whether or not to commit to buying the new and expensive mRNA Covid-19 vaccines for COVAX. "I think the thing we feared was this: we had limited resources, and if we spent a lot of money on expensive mRNA vaccines that we didn't know how or whether we were going to be able to deliver, we'd end up with COVAX having only limited quantities of vaccine and not enough to fulfil our promises." Vaccines made by the Serum Institute of India were priced at just $3 per dose for COVAX. But as it turned out, relying largely on Serum – a cheaper but single manufacturer in a large country – left COVAX in precisely the situation it had feared and sought to avoid. "So the reason we failed was that we were building COVAX in real time. We were fund-raising and procuring in real time and simultaneously. And the funding didn't come fast enough to prevent us from having to make choices based on economising. And because of our fear of the resource constraints, we decided not to take risks that in hindsight we should have taken."

As for Seth – well, coming, as he does, from a youth spent trekking the world as an epidemiologist and an ad hoc expedition doctor – as well as from being the founder and former chief executive officer of the International AIDS Vaccine Initiative – he is no stranger to risk-taking and failure. Seth tells of a time when he was on what he describes as one of his "most exciting" trips. It was a team expedition with ultra-running and trekking enthusiasts to find out whether the Fish River Canyon in Namibia, the second largest canyon in the world at 550 metres deep, would be runable in rainy season. "My father had died only about a week before the trip, and I probably shouldn't have gone. But I thought it would help to clear my head and my sorrow," he told me. "But evidently, I wasn't being as attentive to the terrain as I should have been – and I ended up twisting one ankle and then getting my other leg caught between two rocks and then having a slow twisting fall." The trip doctor's self-diagnosis was that he'd suffered a spiral fracture of one ankle – "it was hanging off" – and torn all the ligaments of the other. Being 400 kilometres or so from medical help, he had to set his own ankle using two reporters notebooks and some duct tape borrowed from one of the other expedition team members. Asked whether this painful experience had done anything to tame his appetite for risk, he admits that six months in a wheelchair forced him to slow down a bit – and put paid to his love of marathon running. But in some ways the experience taught him more about seizing the moment to try to do the right thing with the best of whatever resources you have available. Seth's several decades in global health, seeking out new ways to protect and fight against deadly infectious diseases, have underlined for him, too, that shooting for the moon, even if you only make it to the stars, is certainly the right approach. "The unavoidable fact is that innovation means taking risk," he says. "We must be willing to try new things, but with that we must also be willing to fail, learn from what happened and take those learnings forward."

In the context of the Covid-19 pandemic, it's useful to imagine what might have happened if there had been none of that willingness to try new things. No scientists willing to create new types of vaccines with messenger RNA technology that had never before been proven successful. No global health funders willing to put money behind vaccine research and development projects that might fail, but also might generate billions of affordable doses to help vaccinate the world. And no global vaccine sharing system like COVAX. As of the end of May 2022, COVAX had failed to meet its original ambition of getting 2 billion Covid-19 vaccine doses to participating countries by the end of 2021. But it had delivered more than 1.5 billion doses, with 90 per cent of those going to people living in the 92 poorer countries that would otherwise have struggled to get access to any Covid-19 vaccines at all.

It's because the world tried to design a global vaccine sharing system that would help protect people as fairly and evenly as possible that we have learned what can happen when sharing fails.

Importantly, the very existence of COVAX, with all its successes and failures, has focused the world's attention more than ever before on the absolute need for defences against infectious diseases to be distributed equally. It has brought new phrases, like "vaccine nationalism" and "vaccine equity," into common usage in a way that has forced the world to consider which path is the best one – both for the common good, and for self-interest. It has been revelatory of a global health security system that is not currently configured to succeed in protecting everybody, everywhere – but also revelatory of the ways in which that system can be changed to be more successful in the future. Most notably by being prepared ahead of time. The changes will mean encouraging stronger political leadership when it comes to pandemic prevention, and crucially, creating a pandemic prevention financing fund. That will ensure there is predictable and sustainable cash put up front for the scientific research and development needed to build a library of prototype vaccines. The changes will also mean building a truly global vaccine-making industry to be able to manufacture top quality, low-cost, high-volume vaccines at speed in the event of a newly-emerging disease epidemic, so that in future health crises, no region of the world ever has to rely solely on another to be able to protect its people from disease.

7: PREPARE to Spend Money

One thing we have found out during the Covid-19 crisis is that if disease epidemics are expensive, then disease pandemics are orders of magnitude more expensive. By the end of 2025, the emergence of the novel SARS-CoV-2 coronavirus, and all that it brought with it, is estimated to have cost the world $28 trillion. That's twenty eight thousand billion dollars, or $28,000,000,000,000, wasted over five years.

Of course, there is a range of such estimates. It spans from as low as $15 trillion to as high as $50 trillion. And such estimates are just that: informed, calculated guesses at the amount of money spent and the amount of economic growth and prosperity forgone as the microscopic viruses that cause Covid-19 infected, isolated and impoverished billions of the world's people. But whichever end of the range you go for, it's an almost unimaginable amount of money to the average person. And it's also way, way higher than any pre-Covid predictions went.

In September 2019 – just three months before the novel coronavirus made its entrance – a panel of global health experts known as the Global Preparedness Monitoring Board published a report entitled

"A World at Risk."[38] The board is co-chaired by the formidable former Prime Minister of Norway and former Director General of the World Health Organization, Gro Harlem Brundtland. Brundtland, who narrowly escaped assassination by the Norwegian mass killer Anders Breivik in the summer of 2011, also co-founded an independent group of experienced and wise former global leaders known as The Elders with the late Nelson Mandela. The Global Preparedness Monitoring Board's report cautioned that a global flu pandemic on a scale of and with the virulence of the 1918 Spanish Flu could cost today's global economy $3 trillion, or up to 4.8 per cent of global gross domestic product (GDP). For even a moderately virulent flu pandemic, the predicted cost would be around $1.5 trillion or 2.2 per cent of GDP, the report said. Giving recent historical comparisons for context, it noted that the SARS outbreak of 2003 had cost the world $40 billion, while the 2009 H1N1 swine flu pandemic cost $55 billion. And Zika, a mosquito-borne disease that re-emerged in Brazil in 2016 and spread internationally, cost the Americas something in the region of $20 billion, including millions of dollars of costs in extra care for the disabled children it left in its wake. For Ebola, the best estimate for the 2014 to 2016 epidemic in West Africa is that it cost the world $53 billion on top of the direct impact it also had on the economies and livelihoods in the countries affected. Conservative estimates of the economic cost of pandemics of novel types of flu and other respiratory infections are around $80 billion a year when averaged over a century.

Looking back at the Global Preparedness Monitoring Board's 2019 report now – as I did often during 2020 and 2021 when I reported on the coronavirus pandemic playing out – it seems to me that the most appropriate response is: if only. If only the world's leaders, the finance ministers of wealthy governments, the heads of trillion-dollar corporations and the trustees of billion-dollar philanthropies had read that report and heeded its warnings. If only they had taken it seriously. It's not as though they would have had to read between the lines to figure out what the experts were saying. It was crystal clear. Laid out in black and white. Even, sometimes, in bold type – just to hammer it home. "The world is not prepared for a fast-moving, virulent respiratory pathogen pandemic," it said. "The 1918 global influenza pandemic sickened one third of the world population and killed as many as 50 million people. If a similar contagion occurred today with a population four times larger and travel times anywhere in the world less than 36 hours, 50 to 80 million people could perish. In addition to tragic levels of mortality, such a pandemic could cause panic, destabilize national security and seriously impact the global economy and trade." It went on to say that most countries would need to spend on average between $1 and $2 per person "to reach an acceptable level of pandemic preparedness." And it gave a stark warning: "Not investing is a high-risk gamble," it said. It was indeed.

But then, just as it largely ignored warnings about the biological and social threat of the new coronavirus, the world didn't seem to want to hear this or other economic and financial warnings, even in the very midst of the most expensive pandemic ever to hit it. Another mid-pandemic crystal-clear caution came from the International Chamber of Commerce in January 2021. It warned that fully vaccinating the populations of rich countries while neglecting poor ones could cost rich countries as much as $4.5 trillion in lost economic activity. It said that if rich countries acted fast, and if they invested just over $270 billion into the World Health Organization's accelerator scheme for more globally equitable access to vaccines, tests and treatments for Covid-19, they would avert losses in their own economies of 166 times that amount.[39] This was a "major investment opportunity," the ICC report said. It was indeed.

Given the horrendous humanitarian, political and economic shock the world has just been through, and is still suffering from, we have to begin thinking more seriously about global security, and about how crucial a factor health is in protecting that security. Or, more accurately, how crucial a factor a lack of epidemic disease is in protecting that security. The whole global economy has tanked because we were not as ready for Covid-19 as we should have been, and as we were told we could have been if we had spent just a couple of dollars per person.

Most governments would not think twice about spending money on extra fighter jets, another drone, a few more tanks, or more sophisticated and powerful artillery to beef up their armed forces if they felt even a whisper of an increased threat from a neighbour or a far away potential foe. Most of the time too, those governments are pretty happy with the idea that those extra weapons or defences won't be used – in fact they're hoping they won't. To be sure, spending on national security – often called defence or military spending – is regarded by most people and their governments as a must. One of the first duties of government is to protect the security of its citizens, its institutions and its economy by building and maintaining national security. If we want peace, we must prepare for war.

Which is why we spend so much on it. According to data from the World Population Review in 2020 – the first year in which the SARS-CoV-2 virus engulfed human activity – total world military defence spending was almost $2 trillion, $778 billion of which was spent by the United States alone. China, with the next largest defence budget, spent $252 billion, followed by India at $72.9 billion, Russia at $61.7 billion and the United Kingdom at $59 billion. Since a good chunk of that military spending goes on fighter jets, and because they are a tangible asset that people across the world can imagine in their mind's eye, it's become customary among global health security experts to use these weapons as a cost comparison. In an analysis published in December 2020 about how the world might put together a large-scale plan to develop vaccine defences against a potential Disease X future pandemic,[40] Florian Krammer, an Austrian-American virology professor and vaccine scientist who works at the Icahn School of Medicine at Mount Sinai in New York, chided that when the United States finally decided it would commit funds to CEPI, it did so with an amount less than it would need to spend on buying a single fighter jet. In actual fact, the United States' first funding to CEPI was a donation of $20 million in October 2020, and you could barely buy more than a wing of a F-35 fighter jet for that. The aircraft are currently priced at around $80 million apiece, and the U.S. has thousands of them as part of its combined military aircraft count of 13,247 planes. For the record, the United Kingdom already also has more than 20 F-35s and has said it wants eventually to take that number up above 135, while Canada announced in March 2022 that it hoped to buy 88 of them. A few weeks earlier, Germany's new government, led by Chancellor Olaf Scholz, said it was aiming to buy 35 F-35 fighter jets to replace its ageing Tornados as part of a 100 billion Euro German military upgrade programme.[41]

There's no question then, that when governments want to find money to spend on military defences, it's clear that they can, and they do. Somehow, however, our governments don't yet seem able to take the same protective approach to defending our economic, physiological, mental and societal health against deadly diseases. Richard calls this a critical categorisation issue. We are misunderstanding the nature of pandemic threats. Getting it wrong has put us where we are now – having spent trillions of dollars on dealing with epidemic after epidemic, pandemic after pandemic, instead of bolstering our health security to minimise or neutralise their impact. Getting it right from now on in is fundamental to being able to prepare better for the future. If we view the threat of pandemics as a health or development problem, and we have to "scrounge around down the back of the couch," as Richard puts it, to find money from those much smaller national health or overseas development budgets, then we're going to be scratching and scrounging forever, mostly in vain, to get to the level of investment needed. But if we view pandemic threats as a national and international security problem on a par with the threat of military conflict – as all the evidence from the Covid-19 pandemic suggests they are – then they become part of a package of security issues that simply have to be dealt with, and for which money is largely already set aside. The fundamental problem is, according to Richard, that the world has been making a category mistake. We have been misframing infectious disease threats as a health or development problem. And we can't find a solution if we don't understand and correctly frame the problem we're trying to solve.

Solving the world's biggest problems was what the scientist and writer Dr Rowan Hooper was puzzling over in 2019, when he came up with an idea for a thought- (and hopefully action-) provoking book: *How to Spend a Trillion Dollars: The 10 Global Problems We Can Actually Fix*.[42] Hooper is a London-based editor and writer for the *New Scientist* magazine and the host of its weekly podcast. Chatting on a video call about what he called his Project Trillion, he told me that he'd chosen a trillion dollars in part because it was a "nice round number" and in part because he wanted to inspire people to analyse the world's problems by imagining they really could have the kind of resources they'd need to solve them. A trillion dollars also happens to be roughly one per cent of current global GDP. "I wanted an amount I could really do something with," he said. "And there is actually a huge amount of money sloshing around out there." At the beginning of 2020, just as the novel coronavirus was emerging, private equity firms held $1.45 trillion in what is known in the trade as "dry powder," Hooper explained. That's just cash – not money tied up in assets like houses or stocks – but ready cash, itching to be invested.

Personal fortunes are plentiful too. The world's richest one per cent of people together own an astonishing $162 trillion. As if to emphasise that last point, Elon Musk, the tech entrepreneur, Tesla CEO and the world's richest person, decided in April 2022 that he'd be happy to splash out $44 billion dollars in one go to buy the social media platform Twitter. It was worth it, he said, because Twitter was the place "where matters vital to the future of humanity are debated."

Hooper's Project Trillion rules exclude spending the money on media, politics, investments, the military or just about anything that could entail malicious intent. His thought experiment is to ask people to think how they would spend a trillion dollars if they had one year to do it for the good of the world. The money has to be spent for the good of the planet, for the good of its people, or for the advancement of science to progress either or both of those goods.

Although Hooper's rules ban spending on anything military, it's somewhat galling to note that a trillion dollars is only around half of what the combined world's governments spend each year on military defence and security. It's also less than half of the market capitalisation valuation of corporate giants like Google, Amazon and Microsoft, and a tiny fraction of what governments around the world have come up with to spend on dealing with the emergence of the novel coronavirus and all its deadly and destructive consequences. In 2020 alone, somewhere between $9 trillion and $12 trillion was found – or created through a curious mechanism called quantitative easing, which is essentially when central banks simply print more money – to tackle the global coronavirus crisis. In the first half of 2020, the G20 group of wealthy nations agreed a $5 trillion economic stimulus plan and the European Union passed a €1 trillion rescue package for its members' economies. In the United States, lawmakers in Congress passed an economic stimulus package worth $2.2 trillion to try and head off the worst economic effects of the Covid-19 pandemic.

So while in some ways, a trillion dollars is a vast sum, in others, it's not at all an unreasonable amount to think about spending on preparing to prevent another catastrophe like the Covid-19 pandemic. In fact, Rowan Hooper's aspirational global problem-solving plans extend well beyond preventing future pandemics. He would aim to cure all disease if he had a trillion dollars and a year to spend it, he writes – and he'd start by curing and eradicating HIV/AIDS, doing the same for tuberculosis, and eradicating malaria and other tropical diseases. For vaccine development against emerging infectious diseases like Disease X, and for expanded immunisation programmes to keep current infectious disease epidemics under control, Hooper allocates $100 billion – just 10 per cent of the amount he's allowed himself to play with as he goes about theoretically solving humanity's greatest problems. Yet if we do this right, we might not even need to spend that whole 10 per cent of a trillion dollars. We might even get some change.

Among the many lessons to be learned from the experience of facing the emergence of the SARS-CoV-2 virus is that spending sustained amounts of money over years in advance on research and development into a whole range of potential next Disease Xs makes sound economic sense. Let's go back a decade to 2012.

The kids were in bed, so I was sitting on the sofa doodling around on my work laptop in our south London townhouse on a Sunday evening in September of that year when I saw something worrying pop up on the World Health Organization's "global alert and response" page. A man from the Arab Persian Gulf state of Qatar had been infected with a previously unknown virus and was now critically ill in hospital in Britain, it said. The virus was totally new, but was related to the deadly SARS virus that had emerged in China and caused an epidemic 10 years earlier, the WHO's statement went on. The man from Qatar, who had also recently been in Saudi Arabia, had gone to the doctor on 3rd September 2012, suffering from the symptoms of an acute respiratory infection. On 7th September, he had been admitted to an intensive care unit in Doha, and by 11th September, a decision was taken that he should be transferred to Britain by air ambulance. This was my first real-time journalistic encounter with the emergence of a novel coronavirus. Initially referred to as nCoV, this coronavirus was later named Middle East Respiratory Syndrome, or MERS.

What with it being a headline-grabbing new SARS-like virus, and what with my being a global health and science correspondent for a major international news agency, MERS kept me pretty busy in the latter months of 2012 and, on and off, throughout the first half of 2013. I remained in equal measure intrigued by it and fearful of it as I kept tabs on it over subsequent years. In 2015, I wrote a piece questioning why, even though MERS for almost two years had been viewed by infectious disease specialists as one of the top potential pandemic threats, no one had yet developed a vaccine to prevent it. The article featured an interview with Professor Adrian Hill, a flame-haired Irish vaccinologist who in 2005 became the director of the Jenner Institute at Oxford University. Hill vented his frustration at the lack of a MERS vaccine having allowed a large and very deadly outbreak of MERS to take hold in South Korea three years or more since the new virus was first identified. "How long are we going to wait around and just follow these outbreaks before we get serious about making vaccines?" he said. The South Korea outbreak – which had started on 20th May 2015 with the first case reported in a 68-year-old man who had recently been in the Middle East – spread to a total of 186 people between May and July. It killed 38 of them – a case-fatality rate of more than 20 per cent. The outbreak was subsequently found to have been caused by the importation of a single case in the man coming home from the Arabian Peninsula. He had visited the doctor with a nasty cough, and was treated in several different clinics – unwittingly infecting more than 25 people along the way – before he was diagnosed with MERS. Professor Hill was cross. With very affordable amounts of money invested in MERS vaccine research in good time, the outbreak could well have been prevented altogether.

"Should we have made a MERS vaccine? Yes. Could anyone have afforded it? Yes, the government of Saudi Arabia. So should something be done? Yes, someone should go and develop a MERS vaccine sooner rather than later," he said. Yet it was another four years before Professor Hill and his team at the Jenner Institute finally got the funding they needed to launch an early-stage clinical trial of their experimental MERS vaccine in Saudi Arabia. And the funding came from Britain via the UK Vaccine Network, an aid programme to develop vaccines for diseases with epidemic potential in low and middle-income countries; from the UK National Institute of Health Research through the Oxford Biomedical Research Centre; and from CEPI, which provided money to extend the period for follow up of participants in the trial to 12 months.

Almost 10 years on, MERS is still out there, slowly picking off one victim after another in a barely noticeable but ongoing progression. Over the course of a decade it has infected at least 2,600 people and killed 935 of them – a case-fatality rate of 36 per cent – with the vast majority of cases and deaths in Saudi Arabia. Sadly, especially now in the context of the millions who have died from Covid-19, those 935 deaths and almost 2,600 cases seem such small numbers that it would be legitimate to question how much money it's worth investing in researching and developing defences that might stop them from growing. But even though the MERS numbers were relatively small, CEPI and Richard needed no convincing. Starting almost immediately after its launch in 2017, CEPI began seeking out scientists who could work on MERS, and ultimately signed agreements for $140 million worth of MERS vaccine research.

"What we had at the time was a bunch of sporadic cases and a few large outbreaks – the one in South Korea and another in Riyadh a year or so earlier which had infected around 300 people," says Richard. "People looked at that situation and questioned – why would you want to go to the expense of developing a MERS vaccine?"

That's a perfectly legitimate question if the case for a vaccine, or no vaccine, is framed purely in terms of cases, deaths and infectiousness, or likely contagion, of a disease. But what's missing from that framing is the bit that transforms MERS from a not-much-to-worry-about, ticking-along type of new disease, into one that should demand huge attention and equally sizeable investment. Part of that is the staggering economic impact it had. While the 2015 South Korean MERS outbreak infected only fewer than 200 people and lasted only around two months, it also led to the closure of 2,700 schools and the quarantining of 17,000 people. Health researchers estimated that the total cost of this tiny outbreak, including losses in retail sales and tourism, approached a staggering $10 billion – working out at a cost of more than $50 million per case. A separate study published in 2019 which aimed to help researchers try and estimate the costs to Saudi Arabia of its thousands of cases of MERS over the years found that even just the direct medical costs of caring for and managing a single MERS patient were on average almost $13,000 per case. Some cases, presumably in people who spent longer in hospital and needed more intensive care, racked up costs of $76,000.[43]

MERS doesn't spread between people as effectively as its earlier-born cousin SARS did in 2003, thanks largely to the fact that the virus that causes it needs to be able to get deep into the lungs to latch on to proteins there and start an infection. This is also why cases of MERS infection carry a higher risk of severe illness or death: a virus that gets deep into the respiratory system can do more damage. But the other part of what makes MERS more than just a slowly ticking-along-type new infection lies in its status as a new and still circulating member of the coronavirus family which demonstrates that family's capability to threaten humans. When SARS petered out swiftly after its outbreak in 2003, MERS became the poster child – the prototype – for human-adapted coronaviruses that can spread, kill and pose a pandemic risk.

Which is why, starting from March 2018, barely a year after CEPI itself had come into being, it began to sign off agreements for tens of millions of dollars to be given to scientists working on MERS vaccine development. The first of these was a $37.5 million grant to the Austrian company Themis Bioscience to fund work on vaccines against both MERS and Lassa fever, an acute viral disease that causes frequent outbreaks of haemorrhagic fever in West Africa. The following month CEPI agreed to give up to $56 million to the American biotechnology company

Inovio Pharmaceuticals to work on a DNA-based MERS vaccine; and in August and September it signed off a further $36 million and $19 million respectively to a German-based company called IDT Biologika and to a joint project between Oxford University's Jenner Institute and Janssen Vaccines, a unit of the U.S. pharmaceutical giant Johnson & Johnson. All of this meant that when the mysterious cluster of pneumonia cases in Wuhan was confirmed in the first week of January 2020 as being caused by a new type of coronavirus, some of the world's scientists and vaccine developers had a sizeable head start.

Because money buys scientific knowhow, it also buys speed. Dame Sarah Gilbert's team at Oxford, for example, had used the funds granted to them to study in detail how to design a vaccine against the coronavirus that causes MERS, including how the gene coding for the now infamous spike protein could be plugged into their ChAdOx1 rapid-response vaccine platform. Even before the genetic sequence for the novel Wuhan coronavirus had been published on 11th January 2020, Gilbert's team knew that, because it was a coronavirus, the design of a vaccine against it would be exactly the same as for MERS. And what gave them even more of a head start in developing what ultimately became the Oxford-AstraZeneca Covid-19 vaccine, was that their experimental MERS vaccine had already been through two clinical trials and proved both safe and able to generate an immune response. It was thanks to all this preparatory work, the tens of millions of dollars of research and development into MERS, that the Covid-19 vaccines made to help defend us from the novel coronavirus were able to be developed and brought to the world at Pandemic Speed. In the MERS example, we've seen what the potential payoff can be. CEPI's much-questioned MERS investment agreements – which so far have amounted to $140 million – meant the world had a huge head start against Covid-19. Looking back at it now, the question that should burn in our brains is not why on earth anyone would have spent that kind of money on what seemed at the time like a relatively minor cause of disease outbreaks, but why on earth didn't we similarly spend hundreds of millions more?

While we had bought ourselves something of a head start on coronaviruses, however, the reality is that we are nowhere near as well prepared for a Disease X if it were to come from any other of the 25 or so viral families that we know already have the ability to infect humans. To do that – to cover more of the potential Disease X-producing viral bases – we need to do, at the very least, as much discovery and development research work as we did for coronaviruses with MERS on something in the region of 100 viral pathogens from those 25 families. This will, without a doubt, entail a vast amount of money. And we have to be prepared to accept that much of the resulting work could ultimately go unused. In fact, we'd want to hope that if Disease Xs crop up relatively rarely, and if most of those families may never throw up anything with acute pandemic potential, much of it will go unused forever.

To get the world to a state of preparedness for the next Disease X that would match where we were with coronaviruses when SARS-CoV-2 emerged, virus experts and vaccine research teams will need, for a start, to find out as much as they did about other prototype viruses as they'd found out about MERS. They need to identify and understand the equivalent of Covid's and MERS's spike protein for every other one of the 25 families of viruses, and for subsets of those families if the target proteins are different. Then, they must design and create several potential vaccine candidates for each of those and put them to the test through early stage – Phase 1 – clinical trials to see if they are safe in people, followed by mid-stage – Phase 2 – trials to see if they produce the kind of immune response that might make them effective if there were to be fully developed into deployable vaccines. Such discovery work is time-consuming and extremely costly. And, most importantly, it often fails multiple times before it succeeds. But its end point would be a vast library of candidate vaccines covering all the viruses we now know could infect people. Since any one of these virus families could one day produce something that spills over or mutates to spawn the next pandemic-potential-carrying Disease X, we need to be bold and invest in vaccine candidates against all of them. That will ensure we're ready to move swiftly – at pandemic-busting speed – to adapt and finesse them against any emerging disease threat.

Florian Krammer, the vaccine scientist at New York's Icahn School of Medicine at Mount Sinai, reckons this kind of preparatory work – including the essential scientific research and the two phases of clinical trials – would work out at something in the region of $20 million to $30 million for each successful vaccine candidate. In other words, at the very least $3 billion in total – and that's only with a 100 per cent success rate if none of the vaccine candidates falls by the wayside. But, as we know, the field of vaccine research and development is littered with failed candidates. So it will be many billions more than that if – as expected – many of them do fail. Accounting more realistically for the probability of multiple failures among the initial vaccine candidates, a study published in *The Lancet* medical journal in 2018 concluded that it would take between $319 million and $469 million to take get a single vaccine candidate ready, from initial concept and design to successful Phase 2 trials.[44] Using these estimates, a 100-strong candidate vaccine library would cost at least $30 billion, and could cost up to $50 billion. So, about the same as Elon Musk was prepared to spend on buying Twitter. But unlike Musk's description of Twitter as the place "where matters vital to the future of humanity are debated," the vaccine library would be the place *where matters vital to the future of humanity are created.*

An infinitely more worthwhile investment.

The idea of the vaccine prototype library is that when a future Disease X emerges, the closest-fitting vaccine candidate together with vast files of data and knowledge about it and about the virus it was designed against, can be pulled off the shelf and brought into action. If the newly emerging virus – the next Disease X – is not an exact match for any of the prototypes, a candidate vaccine from the same viral family can be adapted to it and then accelerated into development. Either way, the world would have a huge head start. A $50 billion head start. This is the start of the 100 Days mission that would deliver safe and effective vaccines against any new viral threat in time to snuff out its potential to cause a pandemic.

But we can't get to and make use of that huge head start if the next Disease X is not spotted, identified and communicated around the world in the first place. Which is where the next big chunk of spending comes in. Surveillance. Just as it is in national security, keeping an eye out for enemies – be they old or new, known or unknown – is essential work in health security. Among the millions of viruses that circulate in animals we already know there are at least 260 that can infect people. These types of viruses are known as zoonoses, and they're the ones we need to be particularly on top of. To keep tabs on those, and to do the research to find out how many more potential animal to human viral threats are out there, is expensive work. The USAID PREDICT project which ran for 10 years between 2009 and 2019 cost $238 million and only covered 35 countries, but picked up and identified around 1,200 viruses with the potential to cause human disease. Its successor, the DEEP VZN project launched in 2021, is a five-year plan to explore and assess those threats in more detail, and its budget is $125 million.

The international scientists behind the Global Virome Project, an ambitious plan to detect and identify almost all the world's viral zoonotic risks to human health security, say they need a minimum of $1.2 billion over 10 years to be able to discover around 70 per cent of zoonotic viral threats to humans, and up to $3.7 billion to find and categorise virtually all of them. It's a lot of money, but even at the higher end of that range, it would mean spending around $400 million a year – budget dust for the likes of the G20 wealthy nations.

At the heart of projects like the Global Virome Project is the phenomenon of "spillover" – the ever-present procession of animal viruses into human populations. It was this that was the starting point for so many of the disease epidemics that have plagued humanity over the past century – AIDS, SARS, Ebola, MERS. Remember that around 75 per cent of emerging infectious diseases in humans originate from animals. The human immunodeficiency virus that causes AIDS came from chimpanzees and is thought to have jumped the species barrier when bushmeat hunters shot and butchered a chimp. The spillover of the MERS coronavirus is thought to have come from bats via camels and into people before it was then spread half way across the world to wreak death and economic shock on South Korea. Its earlier coronavirus cousin SARS used civet cats in China on its spillover journey that then took it into 30 countries to cause 800 deaths. And there are plenty more lurking in the places that these came from. Specialists in virology – colloquially known as virus hunters – estimate there could be more than a million as yet undiscovered viruses currently circulating in wildlife. Any or all of these has the potential to jump species and pick up the traits and mutations needed to infect human hosts.

A major new study analysing the cost-benefit of this kind of surveillance in preventing pandemics was published in February 2022.[45] In it, researchers estimated that based on the planet's current population of almost 8 billion people, and on what past pandemics have done to us, we can expect, on average, 3.3 million deaths from zoonotic diseases each year from now on into the future. The researchers also delved into the unpleasant world of figuring out what a human life is worth, in economic terms, in various regions of the world. Sadly, that value is dependent on the wealth of the country a person lives in – leading to a U.S. life being valued at $10 million by the American government's Environmental Protection Agency, the life of a blue-collar working man in India being valued at just under $600,000, and an estimated value of a life in the Central African Republic being put at just under $2,000. Taking this sickening range into account, the economists on the research team calculated that 3.3 million deaths from zoonotic diseases translates to between $350 billion and $21 trillion in economic losses each year. Preventing just 10 per cent of those deaths would not only save those lives, but could also avert up to $2 trillion in economic losses. And what would that cost? Around $20 billion a year globally spent on "primary prevention actions," says Aaron Bernstein, a Harvard professor and one of the lead authors of the study. In other words, around five per cent of the lowest estimated value of the lives already being lost to emerging infectious diseases every year. Bernstein says this presents the world with an "abundantly clear" spending choice: "The wisest dollar spent on pandemics is the one dollar spent to make sure they never start in the first place."

The Global Preparedness Monitoring Board's reports are annual, which means that after hammering home such piercing warnings in 2019, the panel of experts got their chance the following year to say "we told you so." Graciously, Brundtland and her fellow board members opted not to exploit this opportunity too much. Yet the GPMB 2020 report did loudly lament what the world had let happen to itself. Entitled "A World in Disorder," the 2020 report sought, once again, to cajole, inspire and persuade world leaders to reframe and recalculate their thinking about spending on stopping the next Disease X before it becomes a pandemic.[46] "The return on investment for global health security is immense," it said. "It would take 500 years to spend as much on investing in preparedness as the world is losing due to Covid-19."

Investment in public health prevention is always a tough sell. While it's relatively easier to get governments to spend big money, fast, on a deadly, real, in-your-face-right-now global disease outbreak, it's remarkably hard to persuade them to fund the global surveillance and detailed research and development needed to prevent one. And that's even now – when the evidence could not be more obvious and the warnings could not be more stark that this would save unfathomable amounts of money and millions of lives. That's because effectively, we're asking governments to invest in creating a system that will bring about a non-event. A system that will ensure nothing really bad happens. No global health crisis. No pandemic. No deadly drama. No economic catastrophe. An expensive calm. Just like the expensive peace our governments seek to buy with trillions of dollars of military spending. Every year, treasuries around the world cough up billions of dollars to keep fleets of expensive fighter planes and nuclear-armed submarines permanently patrolling the skies and the oceans to protect us from a threat that may never materialise.

The big question is how much are we willing to pay to prevent another Covid-19? To prevent an event that kills tens of millions of people and costs the global economy trillions of dollars. To create a pandemic-free world. Looking just at the trillions that Covid-19 cost the global economy – let alone at the lives it ended and crushed, and the futures it stole – spending at least $80 billion, and up to $100 billion, on preparing to avoid anything like that again seems like a pretty good investment. "There's a premium that the world must pay to be prepared to prevent pandemics," Richard says. "That's the price of living in the 21st Century."

8: PREPARE for the Next One ...

... because there will be a next Disease X, without a doubt.

While the 1918 Spanish Flu holds the record for the deadliest pandemic in human history, and while Covid-19 may be the deadliest pandemic the world has seen in the 100 years since then, there's nothing at all to suggest we're now due a period of respite from viral attacks. Indeed the data suggest quite the contrary. A major new study published by scientists at the University of Padua in Italy and at Duke University in the United States in the summer of 2021 underlined the message. Far from being once-in-a century freak events, infectious disease epidemics with pandemic potential are nowhere near as rare as we might think. Or as we might hope. After assembling and analysing a global dataset of epidemics of killer pathogens such as plague, typhus, smallpox and multiple types of flu spanning the last 400 years, the researchers found that the probability of a pandemic with a similar potential impact to Covid-19 is now about two per cent in any year.[47] This means the probability of the world facing a pandemic threat on a similar scale to Covid-19 in our lifetime is about 38 per cent. That's not quite a fifty-fifty chance, but it's uncomfortably close.

And far from receding as the Covid pandemic begins to fade, that risk is growing. As the world's population continues to rise, and as more people engage in international trade and travel, and human development increasingly means encroaching on the previously wild territories of animals and birds, it is becoming more likely by the day that once-containable local or regional infectious disease outbreaks risk exploding into global health security threats. The Padua-Duke study found that the annual probability of extreme epidemics occurring could increase threefold in coming decades. Based on the ever-increasing rate at which previously unknown pathogens – Disease Xs – have broken loose in human populations in the past five decades, the researchers estimated that a pandemic similar in scale to Covid-19 was likely within 59 years – a number they described as "much lower than intuitively expected." Another Disease X with pandemic potential is coming. We need to be ready for it.

Among the very few potential positives to emerge from our devastating encounter with Covid-19, the Disease X of the 2020s, is that it can help us identify the traits and skill sets we need to hone to be able to make future disease outbreaks less likely to be catastrophic for the world and its people. We now know that we need to be scared – suitably so – of things that we may not be inherently programmed to be frightened of. Things we can't see and have never seen before. Pathogenic threats. Unlike our in-built human fears of heights or fires or explosions, being aware of, and afraid of, becoming infected with something that is microscopically tiny, invisible and silent is something new. Something SARS-CoV-2 has taught us.

We now know how, using agility and collaboration, we can alert the world to an impending threat in time for its national and international leaders to do something about it. To act swiftly and decisively, if they choose to.

We now know that if those leaders are swift and bold in their actions, we can implement effective social and economic measures that can dramatically slow the spread of an infectious disease. And we know how, using brave investments and risk-taking minds, we can harness the power of science to develop effective new rapid-response tests, vaccines and medicines against diseases we have never encountered before.

We can, but will we?

There is already some, albeit scant, evidence that we are learning at least a few of the right lessons from the Covid-19 pandemic. International negotiations on the drawing up of a Pandemic Treaty – a process that began in 2022 and is due to be finalised in 2024, kicked off with an acknowledgement that transformative change is needed. The idea is that, in an effort to do better at pandemic preparedness and prevention next time – to break what's often referred to in global health as the "cycle of panic and neglect" when it comes to infectious disease epidemics – governments from across the world will find areas where they commit to taking certain actions in the event of threatening disease outbreak, and be held to those commitments via a legally binding accord. The plan is not to rebuild and shore up the flawed disease defence systems of the past, but to recognise that the emergence and spread of diseases is a global threat and to recognise that such a threat has serious consequences for public health, human lives and economies – in other words to be prepared to be scared. The proposed treaty also reiterates that such ever-present threats call for the widest possible international cooperation: a new world order which puts global health security where it belongs, on a par with national defence against invasion, international security from war, and the fight against climate change.

Those initial pandemic treaty negotiations had a sharpened cognisance that the world's plans to prepare for and prevent pandemics must have equity and accessibility embedded in them from the get-go. The counterproductive vaccine nationalism that both deepened and prolonged the Covid-19 pandemic is something few governments want to be associated with in future. In talks, at least, we are recognising that something has to be done – that we need to be prepared to share – and that this means encouraging and supporting, mainly through at-risk funding of new partnerships or ventures, sustainable and regionally diverse manufacturing capacity in countries where there was previously little or no medical product-making infrastructure.

In practice, there are also a few green shoots of hope. The gaping hole in vaccine-making capability in Africa that was exposed so clearly by the

2009 H1N1 pandemic and again a decade later by the Covid-19 pandemic is beginning to be filled. With that, the dangerous over-reliance of one region on another for vaccines at times of crisis may become a thing of the past. Before the Covid-19 pandemic, 99 per cent of Africa's vaccines were manufactured outside the continent. As well as leaving African countries right at the back of the queue when it came to getting the pandemic vaccines that were so swiftly developed and distributed in wealthier countries, this almost total absence of a vaccine-making infrastructure in Africa has hampered scientific and economic progress on the continent for decades. No more, declared the former head of the Africa Centres for Disease Control John Nkengasong in the summer of 2022: "We need to develop a road map and a framework for Africa's vaccine manufacturing."

Already, this framework is beginning to take shape. In August 2022, a South African drug firm called Aspen Pharmacare signed a deal with the vaccine mass production giant Serum Institute of India to manufacture and distribute four types of vaccines in Africa. A few weeks later in West Africa, Nigeria's Health Minister Dr Osagie Ehanire – who six months earlier had lamented the searing lack of humanity in the rich world's hoarding of vaccines during the Covid-19 pandemic – said his country, too, would join forces with the Serum Institute of India to begin domestic manufacturing of vaccines for Nigeria's routine immunisation programmes. Up until then, Nigeria – Africa's most populous country with 218 million people – had been importing every single one of the millions of doses of vaccines it administered to its people each year to protect them from everything from polio and measles to tuberculosis, pneumonia and yellow fever. Moderna, the U.S. biotechnology company which developed an mRNA vaccine against Covid-19, also decided in mid-2022 to move into Africa, putting $500 million into creating capacity to make mRNA vaccines in Kenya. And BioNTech, the German biotechnology company that teamed up with Pfizer to develop the very first mRNA vaccine against Covid-19, began building a vaccine manufacturing plant in Rwanda, with plans to build another in Senegal.

The hope is that by 2040, 60 per cent of the vaccines used in Africa will also be made in Africa. As well as boosting African self-sufficiency, this expanded vaccine-making capacity should swiftly become a driver for scientific research and development on the continent. But to make these new factories viable and sustainable, to ensure they are still in business when the next Disease X pandemic threat looms, governments will need to continue to commit to buying regular supplies of routine vaccines from them in 'peace time' – in other words when there is no current international disease outbreak emergency. Then, in the event of another Disease X pandemic threat, the expanded capacity will mean that factories all over the continent of Africa should be able to switch to pumping out millions of doses of potential rapid response pandemic-preventing vaccines within weeks of a need for them arising.

There are some signs, too, that the world has acknowledged the need to take risks and spend money if it is effectively to defend itself against a future Disease X pandemic threat. While it's only a modest step, the establishment by the World Bank and the World Health Organization in September 2022 of a facility known as the Financial Intermediary Fund, or FIF, was at least a recognition that having a pool of money ready for quick release surge financing when the next Disease X pandemic threat emerges is essential. The FIF – effectively a trust fund held at the World Bank that pools private and public cash – is designed to provide a $10.5 billion a year slush fund for global preparedness and response financing. There's still a long way to go, however, before we can tick the box to say we are really prepared to spend serious money. At its 2022 launch, the FIF had only around 10 per cent of the $10 billion a year it needs – money pledged by the United States, the European Union and around 20 other national government and philanthropic donors.

For an optimist, these green shoots offer some hope that leaders of governments around the world are learning enough from the devastation of the 2020 coronavirus pandemic not to allow the world to be caught so vulnerable and unprepared ever again. But not everyone is optimistic.

In May 2022, at that year's postponed World Economic Forum gathering of world leaders in Davos in the Swiss Alps, Richard and Seth caught up at an old-style steakhouse for dinner with the American epidemiologist Larry Brilliant – the man who had in the 1970s helped the world conquer smallpox, the worst human disease in history. Despite being a self-described visionary physician and spiritual seeker – labels he assigns himself in the subtitle of his book *Sometimes Brilliant* – the now almost 80-year-old global health expert was not feeling particularly positive about the future. Over steak, French fries and red wine, he offered his companions a sobering assessment of the world's current capabilities in fighting deadly infectious disease epidemics. If smallpox had not been eradicated worldwide in 1980, and were instead still with us, he said, we would not now have the collective capability to overcome it. As a global community we would not be able to come together with the collaboration and determination needed to wipe it out. In essence, Brilliant's assessment was this: post the Covid-19 pandemic, the world's nations are less united, more fractious, and more destructively self-interested than they were even in the 1960s and 1970s, when the arch enemies of America and the Soviet Union came together to lead a successful global effort to wipe out the killer disease despite simultaneously pointing missiles at one another.

It was a grim appraisal of a world whose people would, I suspect, largely think of themselves as more globalised and internationally-minded than their grandparents or great-grandparents could ever have been. Grim, but perhaps unsurprising. With war in Europe seeing Russian tanks rolling into Ukraine, soaring oil and gas prices driving an international inflationary cost of living crisis, and these two factors then combining with climate changes and extreme weather events to create looming global food shortages, we were already by late 2022 at least three or more crises beyond the Covid-19 pandemic, even though the pandemic itself was not over. It feels as though these last few years really have been an era of plague, war and famine. An era of global threats and interlinked emergencies stacking up faster than the world can cope with them. A time when geopolitics is creating an extra layer of problems for the world, just as it is already faced with so many.

In this context, when accumulating crises are sapping our collective resilience, when governments and nations are increasingly ideologically at odds, and when people feel more divided and more distant from each other than they have for a generation, the big risk is that we will quickly slip back into the cycle of panic and neglect when it comes to pandemics. Yes, we were discussing some kinds of solutions – a Pandemic Treaty, a Pandemic Fund, more surveillance and sharing of disease data, more vaccine-making capacity in the Global South – but governments' ambitions can be limited. Their attention can swiftly begin to wander away from yet-to-materialise disease outbreaks that they find it so hard to care about when their more urgent focus is on short-sighted political gains, on how to win votes a few months or a year or two ahead.

Echoing Larry Brilliant's springtime worries in an interview a few months later, Sir John Bell, the Canadian-British Oxford University professor of medicine who has been an advisor to several UK prime ministers on science strategy and who worked on Oxford's Covid-19 vaccine with AstraZeneca, admitted that his hopes were already fading that government leaders around the world were really taking pandemic preparedness and prevention seriously. "I was one of the people who thought the (Covid-19) pandemic would be a great incentive to sort ourselves out," he told Britain's *Telegraph* newspaper. "It hasn't quite worked out like that." Similarly, in its 2021 report, the Global Preparedness Monitoring Board gave another of those warnings – much like the ones I wished the world's political and financial powers had heeded in the 2019 and 2020 GPMB reports. "If we do not change course – even with the results of our failings starting us squarely in the face – we will have squandered a rare and fleeting opportunity to implement the transformative changes needed," it said.[48]

But times of crisis, catastrophe and change are also times when we can make choices. And even in the past 100 years, we've been here before and managed to make some good choices. The devastation wrought by World War II – an event arguably comparable to the Covid-19 pandemic in the enormity of the lives and livelihoods it effected – caused us to re-examine ourselves, to look at the divisions in our societies and between our governments, and to figure out how to build a new international system that could keep us at peace.

So as we emerge from the deadliest infectious disease pandemic in that same past century, we can choose to raise our heads – to prepare to listen

- and learn the lessons from what went horribly wrong, and what went comparatively well. Or we could choose to bury our heads, close our eyes and ears to the lessons, and blunder on, hoping that in putting Covid-19 behind us, we can somehow make the idea of pandemics – and with it the threat of future ones – just go away.

We know deep down, however, that the second option is a no go – a defeatist and defeating cop out. The threat of pandemics is not fading into history. It's looming larger than ever, as all the surveillance and spillover studies show. And while it's tempting to throw up our hands, to be cynical and to think things are never going to change, and that the unfairness and inequalities that doomed us to years of suffering during Covid-19 are inherent in our societies, I'd like to hope we can be less pessimistic. Yes, geopolitical divisions, fractious societies, short-termism and economic downturns are going to make things a lot more difficult. Yet if we allow them to prevent us from building stronger defences against the next Disease X threat, we will only exacerbate them and allow them to persist. Instead we need to create a global environment where each of us – every healthcare worker, every potential patient, every scientist, every vaccine developer and maker, every voter, every politician and every pandemic worrier – can play our part in a creating a global health security ecosystem that does what it's supposed to. Keeps everyone safe.

I'm confident that I'm right in thinking that no-one who lived through the Covid-19 pandemic ever wants to go through something like that again. I'm also pretty sure that I'm right in thinking we don't have to. This book was inspired and shaped not just by my experience of reporting on Covid-19 – a world-changing event that I had heard global public health experts warn about for years – but also on the 2009 H1N1 pandemic, the MERS coronavirus epidemic, the 2014-2016 West Africa Ebola epidemic, on AIDS, tuberculosis and malaria, and on the exceptional progress the world's scientists have made in developing medical defences – specifically vaccines – against infectious diseases both old and new. It was also shaped by my learning just how much the prescient work of pandemic worriers like those at CEPI had been ignored or dismissed as fearmongering, and just how different things could have been if it had not.

The pandemics and epidemics were all international emergencies that had a deadly reach, far beyond just health and medicine – causing economic, political and societal disruption on a scale vastly disproportionate to the size of the pathogens that caused them. The vaccines, while each marking an enormous step forward, all came too slowly and too unequally to save the numbers of lives they could have if they'd been made, and made available, to everyone that needed them more quickly.

But what if it took just 100 days to build effective human defences against any new virus, any mutation of a known viral threat, or any Disease X? Humanity would be one step ahead, and deadly outbreaks could be stopped before they spiralled into pandemics. This book describes the ingredients and skills we need to create a world in which future inevitable disease outbreaks can be contained within the first 100 days of their global threat being recognised, before they get the chance to spread their destruction across the world. If we get the mix right now – the mix of fear, speed, bravery, collaboration, success, humility and investment – and if we keep it up, we can contain Disease X threats and make our future free of pandemics.

9: 2027: A Pandemic is Thwarted

It is June 2027 and doctors at the Sunshine Hospital in the city of Kompang Chom, Champosa, are feeling increasingly helpless and hopeless as they watch Boupha's parents sob at her bedside. Between terrifying fits and seizures, the little girl's glazed eyes peer unseeing into their faces. Her breathing is shallow, and the beating of her weakening heart unconvincing.

The child's health had already been deteriorating when she'd arrived at the provincial hospital. She had been back at school after an extended weekend off for Champosa's National Holiday of Thanksgiving on 6th June. After the morning's lessons, though, Boupha had felt both hot and cold, and headachey too. She didn't feel much like eating the delicious bowl of fried pork with rice and ginger that her father had made for her to take to school for lunch. She went to her teacher, who, seeing Boupha's pale and feverish face, had called her parents to come and take her home. The weekend before, eight-year-old Boupha had been away for a few days off school and out of the city, staying with her aunt, uncle and cousins in a village a few miles away by the river. The little girl and her older cousins, Chantrea and Kong-Kea, had spent a wonderful few days, in the mornings playing in the humid, dappled air of the forest, and in the afternoons dipping in and out of the cool waters of a the stream that led to Pechong River. Now, in Sunshine Hospital's children's ward, those few days of wild-running freedom seemed an age away.

Dr Lita Kim, the paediatric consultant on duty, had immediately prescribed antibiotics for Boupha when she was admitted to the ward the previous evening. A swiftly-delivered intravenous course of a broad-spectrum antibiotic might, with any luck, begin to fight the pneumonia beginning to flood Boupha's lungs with fluid, Dr Kim had reckoned. That assessment, however, had proved wrong. With Boupha's ability to breathe becoming weaker by the hour, Dr Kim's team decided to move her onto oxygen. First, they delivered this via a mask covering her nose and mouth. Then they moved her onto a mechanical ventilator that would take on the job of breathing for her. Dr Kim – an experienced specialist who had seen a whole range of infectious and chronic diseases in thousands of young patients over the years – was by now baffled, and becoming increasingly troubled. With Boupha suffering high fever and aching pain, Dr Kim had also wondered whether the culprit might be Dengue Fever – a viral infection carried and spread by mosquitoes and a disease commonly seen the months after Champosa's May to June rainy season. But a blood test carried out at the hospital's in-house pathology lab had shown no signs of Dengue infection in Boupha's blood. Now, Boupha's worsening symptoms were pointing more towards a respiratory infection of some sort, Dr Kim thought, and one that had gone deep. Not just into Boupha's lungs and heart, but into her head, where Dr Kim suspected encephalitis was beginning to wreak havoc in her brain.

Alarmed by Boupha's rapid decline, Dr Kim went back again to the child's parents to ask for a day-by-day account of where their little girl had been and what she had been doing for the past few weeks. They told her about Boupha's trip to her aunt's and uncle's village house by the river, and about how she had gone back to school after the National Thanksgiving Holiday break, only to be sent home sick by her teacher. Growing more alarmed as she heard this account, Dr Kim asked her team to take down as many names, addresses and phone numbers as they could get of Boupha's relatives, teachers and classmates, and to contact them to ask whether they too might be feeling unwell. She also decided to send a small sample of Boupha's blood, and another one of her sputum, urgently to Champosa's National Institute of Virology, or NIV.

Despite being a relatively small institution with only one main laboratory, Champosa's NIV was well equipped with the latest state-of-the-art gene sequencing technology and had an expert team of specialist scientists on its staff. Its director, Professor Eam Sokhan, was a specialist in emerging viruses and had trained under some of the best infectious disease specialists in China and the United States before returning to her South East Asian homeland after the catastrophic Covid-19 pandemic of 2020 to 2023. Arriving for work on that Wednesday in mid-June 2027, Professor Sokhan received a WhatsApp message from a friend and former colleague at the Sunshine Hospital in Kompang Chom. Dr Kim had been Professor Sokhan's mentor during a crucial year of her clinical training back in the early 2000s and the two had remained in sporadic touch since then. "Have sent you a sample from a paediatric case that is causing me some concern," the message said. "Patient suffering pyrexia, hypoxemia. Some signs of encephalitis. Our lab has not been able to identify agent. Dengue and malaria negative. Would very much appreciate your urgent attention."

Professor Sokhan headed immediately for the lab to see if the samples from Dr Kim's patient had arrived. Finding that they had indeed – in an urgent delivery of a metal box containing test tubes packed in dry ice – she directed her team to begin running tests on them straight away.

Dr Kim's and Professor Sokhan's sense of urgency was something that had been cultivated in Champosa over more than a decade. Like many others in the region, the country had a history of having to deal with imported new and re-emerging infectious disease outbreaks including SARS in 2003, Zika in 2015 and Covid-19 in 2020. Since 2018, all hospitals had been required to report notifiable diseases within 24 hours to a central national database. This was to ensure officials at the Ministry of Health could immediately begin to track epidemiological developments in any part of the country. At around the same time, in 2018, and in collaboration with the Centers for Disease Control and Prevention in the United States, Champosa had implemented a new and innovative "event-based" on-the-ground surveillance system which involved and empowered members of the public including teachers, pharmacists, religious leaders, and even traditional medicine healers, to report any unusual or striking health events. The idea behind this system was that it would help swiftly identify any clusters of people with similar symptoms that might suggest a disease outbreak was emerging.

At the National Institute of Virology that day, what the testing team found did not match anything they had seen before. This was entirely new. The pathogen invading and replicating in Boupha's cells looked a little like a virus called Nipah, an emerging zoonotic disease first identified in people in Malaysia in 1998 and in Singapore in 1999. Nipah causes a brain-damaging infection and, in recent outbreaks in humans, had had a horrifying death rate of up to 80 per cent. Similar to the Nipah virus, and also similar to a related but rarer deadly zoonotic virus called Hendra, this new virus was spherical in shape. The multiple protrusions reaching out from its surface were long, slim and tentacle-like. It also had some features that looked a little like the measles and mumps viruses, both well known to virologists and both previously the cause of decades-long epidemics around the world. This virus was not measles or mumps, and neither was it Nipah or Hendra, but it was almost certainly related to them. Clearly a member of the paramyxovirus viral family, the scientists concluded, this was a virus that had not been seen in humans before.

Paramyxoviruses are frightening. They are known to infect a wide range of animals and birds, including pigs, bats, skunks and even snakes, seals and dolphins. In people, the paramyxovirus family is responsible for highly contagious diseases like measles and mumps which have stalked human populations for centuries. And as well as the lethal Hendra and Nipah viruses, the family's offspring also include several types of common cold and respiratory viruses that, while mostly mild, can spread like wildfire. Adding to their potential to pose human pandemic threats, many paramyxoviruses are respiratory pathogens – a feature that means they can be spread in coughs and sneezes via aerosols and droplets that can hang in the air or contaminate surfaces with viral particles. Zoonotic animal to human outbreaks of both Nipah and Hendra in people had previously been originally sparked by people coming into contact with contaminated faeces from fruit bats.

Concerned to double and triple check what they thought they were seeing, the NIV scientists called Professor Sokhan into the lab. She duly came, bringing with her another batch of samples from Dr Kim's hospital. Boupha's aunt, a 42-year-old agricultural team leader on a nearby rice plantation, had since been brought into hospital with similar symptoms to her niece: cough, fever, sensitivity to light. Boupha's young cousin, too, was now on Ward Four of the Sunshine Hospital being treated for respiratory symptoms and confusion. Testing samples from all three patients – the little girl, her 12-year-old cousin Kong-Kea, and his mother – the NIV scientists kept finding the same thing: a new type of paramyxovirus virus was invading their bodies. And so far, their immune systems were barely managing to mount a defence.

Back at the hospital Dr Kim was not yet aware of what the NIV virologists had found, but she was already taking precautions. She had quickly and quietly decided to put up an isolation zone around poor Boupha and ordered the staff caring for her to put on full protective gear. She did the same when Boupha's aunt, and then her cousin, were admitted. She didn't want to cause a panic, but she was sure that alerting staff to a potential biological threat early would have few downsides, and might prove to be a lifesaving move in the longer run. By mid-morning on the Friday, Boupha's teacher and some of her classmates had been contacted by the hospital's staff. Nurses advised them to stay at home for the time being, monitor their temperatures, be on watch for coughs, joint aches, or any painful sensations behind the eyes, and to call her immediately if any of these appeared.

On Friday lunchtime, Dr Kim got an email from her friend and former colleague Professor Sokhan. The news from the sample testing was extremely worrying, Professor Sokhan said. The pathogen was a virus with strong genetic similarities to other paramyxoviruses, but one that had not been encountered before in humans.

"I think we should alert the Ministry of Health as soon as possible," Professor Sokhan wrote. "I realise there's an awful lot we still don't know about this virus, but it is novel and clearly has potential to be dangerous. It looks very worrying to me."

DAY 1

On 17th June 2027, a member of the World Health Organization's country office in Champosa emailed colleagues at the WHO Pandemic Hub in Berlin, Germany, to alert them to an evolving situation in his country: scientists at Champosa's National Institute for Virology who had been asked to investigate and run tests on samples from a cluster of sick patients at the Sunrise Hospital in Kompang Chom had identified what they believed to be a novel paramyxovirus. Later that night, staffers at the WHO's Geneva headquarters monitoring the agency's Epidemic Intelligence from Open Sources (EIOS) platform picked up a media report from the global disease tracking website ProMED. It relayed preliminary reports of about a half a dozen cases of a "concerning" viral infection of "unknown cause" centered around a hospital and school in a city not far from the Pechong River in Champosa. Immediately extremely concerned, the WHO's emergency response director, Cate Henderson, activated the WHO's Incident Management Support Team and directed them to seek more information urgently from Champosa's Minister of Health and from Professor Sokhan's team at the institute. At the same time, Henderson decided the WHO should issue a notification. The note would go firstly via the WHO's Event Information System, which she knew would be accessible to all of the agency's 194 Member States, and secondly via the Global Outbreak Alert and Response Network, which included several fellow United Nations agencies, major national laboratories, and public health agencies such as the United States Centres for Disease Prevention and Control.

"On 17th June 2027, the WHO's Champosa Country Office was informed of a small number of cases of an atypical viral infection causing severe illness in several children and adults in Kompang Chom," the statement said. "Among 15 cases reported, eight are in children under the age of 12 years. Serious illness has been reported in three of the paediatric cases and severe illness in a further two cases in adults. All patients are at the same hospital and being cared for under precautionary isolation procedures. The causal agent has not yet been confirmed, but preliminary findings of initial testing of patient samples by Champosa's National Institute of Virology point to a possible novel virus of the paramyxovirus family. If confirmed, these results would be very concerning and would warrant urgent action under the terms of the 2024 Pandemic Treaty agreed by Member States. WHO advises Member State governments to actively monitor the situation and alert relevant public health and laboratory facilities."

On Twitter that same day, media reports from the *South Asia Times* and the *Champosa Daily Record* were picked up and retweeted by several thousand users. A mysterious illness appeared to be spreading among people in the city of Kompang Chom, the reports said. The patients, some of them children, were initially diagnosed with viral encephalitis, a kind of brain inflammation caused by an infection. Some were also suffering from pneumonia. The WHO had already signalled its concern after initial virological investigations pointed to a possible novel paramyxovirus. The media reports also noted that Kompang Chom was a busy city with a population of 300,0000 and was on the delta of the Pechong River, a major a channel for regional and international trade, travel and tourism.

20TH JUNE 2027 (DAY 4)

Standing at his desk in the home office he'd been hybrid working from off and on ever since the Covid-19 pandemic of 2020-2023, Aadhan Chakara, a molecular biologist by training and now head of vaccines at Giza Biologicals in Hanoi, Vietnam, fired up his laptop and opened up his browser. Stopping only momentarily to check his email inbox, he quickly navigated to the website of the GISAID database used by fellow scientists around the world to share details of circulating viruses. There, he saw immediately what he'd been looking for: a post uploaded by Professor Eam Sokhan's team at the Champosa National Institute of Virology containing the full genetic sequence of a novel virus that had, the previous day, prompted the WHO to send a Global Outbreak Alert Notification. Across his screen, the hundreds of rows of letters in a plethora of patterns and combinations revealed a unique identity for the newly-discovered pathogen. It was a paramyxovirus, but one never previously seen in people.

With these pages full of code, Chakara was able to get moving straight away. Having worked for the past 10 years on developing a vaccine against the Nipah virus – a pathogen that the WHO, several Big Pharma companies, and a number global health organisations had singled out for special attention as a potential pandemic threat – Chakara was already pretty sure of one or two crucial things about this novel paramyxovirus: that it would more than likely have either a G protein or an F protein – or both – on its surface; that it would use one or both of these as its mode of entry into human cells; and that to design an effective pharmaceutical defence against it, vaccinologists would need to target one or both of those proteins and induce a robust immune response to them.

For this first intense period of work, Chakara didn't need a lab. Just a laptop. He emailed a few members of his team who he knew had extensive expertise on the paramyxovirus family and would have valuable insights on which traits and capabilities this new viral foe might have evolved. Together they set to work designing and tailoring a vaccine candidate.

Over a period of 48 hours from that day in June 2027, this attentive and anxious scene, with individual vaccine designers or teams of two or three developers eagerly downloading the novel virus' genetic code and calculating how it could be plugged into a rapid response vaccine platform, was played out in at least a dozen home offices or labs across the world. In New Zealand and India, separate teams of scientists looked at how they might slot the genetic sequence for the novel virus' G protein into an adenovirus vector-based vaccine design that had proven safe and successful in the AstraZeneca Covid-19 vaccine in the 2020 coronavirus pandemic. In Britain, Bangladesh and South Africa, vaccine developers at two universities and at a biotechnology firm called Artemix Pharmaceuticals also looked a viral vector platform option, but this time using the measles virus – also a paramyxovirus – rather than an adenovirus as the vector. Meanwhile in South Korea and the United States scientists began work on vaccine platforms using a messenger RNA (mRNA) technology that had been developed in the 2000s and then proven successful during the 2020-2023 Covid-19 pandemic.

In each case, the vaccine developers took as their starting point a prototype vaccine from the already large and still growing Global Library of Innovations for Pandemic Prevention, or GLIPP. The library was an online database that vaccine and infectious disease experts around the world had been building over the past five years. Although the GLIPP aimed ultimately to stock at least 100 prototype vaccine constructs covering representative viruses from all of the 25 viral families known to be able to infect humans, in mid-2027, it was still a work in progress. Sections relating some of the 25 viral families – the bornaviruses and the parvoviruses for instance – had only a few pre-clinical vaccine candidates that had yet to be verified in early-stage human testing. In the paramyxovirus section, however, the GLIPP was already fairly well stocked. For measles, for example, a well-known member of the paramyxovirus family, scientists in Australia had in 2025 successfully updated the decades old vaccine and transferred it on to a rapid response platform using a new type of technology known as a molecular clamp. Also in the library's paramyxovirus section, another team in the United States had developed a safe and effective Nipah vaccine based on a platform that used an adenovirus as a vector. For the Hendra paramyxovirus, separate groups of scientists in Bangladesh and the United States had in 2026 proven that an mRNA-based vaccine against Hendra that targeted both the F and G paramyxovirus proteins, and a viral vector-based vaccine that targeted the G protein alone, were both safe, swift to make, and generated protective immune responses in a small-scale clinical trials. Any one of these prototypes, or perhaps all three, could potentially form a basis for a new paramyxovirus vaccine. And any one, or all three, could be manufactured at an accelerated rate of hundreds of millions of doses per month if surge capacity were needed.

Part of the reason that the GLIPP's section on paramyxoviruses was so well stocked with a range of prototypes was that for Nipah, one of the most recently-identified human-adapted paramyxoviruses, international donors had injected a sizeable funding boost for research and development after a global conference on the disease held in Indonesia in 2023. Alarmed by scientists' warnings about Nipah's pandemic potential, government and philanthropic donors had pledged up to $200 million for scientists to speed up the development of rapid response Nipah vaccines. To some at the time, that injection of cash had seemed almost unreasonably large – especially since the several outbreaks of Nipah in previous years, while deadly, had all proved relatively small and ultimately containable. Now, though, with the emergence of a novel virus in the same family as Nipah, the results of that investment were looking like they might show their worth many times over.

For Chakara and his Giza Biologicals colleagues in Hanoi – and for a dozen or so other teams around the world who, like them, leapt on the novel virus' genetic code to try to design a new vaccine – all of this work with measles, Hendra and Nipah was a massive head start. By 22nd June, just six days after the WHO was alerted, media outlets and WHO updates were reporting 45 confirmed cases of infection with the novel paramyxovirus – now being referred to in news reports as the Pechong virus after the river near where it was identified in Champosa – and three deaths. But the outbreak had not, yet, spread beyond Champosa and had not been declared an emergency by international authorities. Nevertheless, by that same date, designs of at least six potentially viable constructs for candidate vaccines had already been created, ready for labs to begin producing the starting material to make them.

23RD JUNE 2027 (DAY 7)

By 23rd June 2027, the WHO was issuing daily situational reports on the novel paramyxovirus to governments, international organisations and the global media. In that day's report, it shared several new developments. In the past 24 hours, eight confirmed and five suspected cases of the novel viral infection had been reported by doctors in the Champosan city of Mon Kuov about 90 kilometres to the south west and down river from Kompang Chom. In this new cluster, so far, only one confirmed case – a 51-year-old construction worker – had died, the WHO said. But in the original Kompang Chom cluster being monitored by the Sunshine Hospital and local public health officials, case numbers had grown rapidly – to 47 confirmed and another 12 suspected infections – and these included healthcare workers at the hospital, it added. A few patients, including some of the infected healthcare workers, were suffering relatively mild symptoms of fever, headache and fatigue. These patients, like all other confirmed and suspected cases, were being cared for respiratory isolation, the WHO said, but did not at this stage require intensive care. The situation was being further complicated by scores of people described in local media reports as "worried well" showing up at emergency departments and health clinics asking to be tested.

In response to the increasing numbers in Kompang Chom, and to the appearance of the new cluster of infections down river in Mon Kuov, local authorities were instructed by Champosa's central government to go into immediate and focused lockdown. Local education and public health officials had already made the decision a few days earlier to advise all schools to close and to switch classes to online lessons for all pupils aged four years and up. Now, though, with chains of transmission showing sustained person-to-person spread in the original Kompang Chom outbreak and a new outbreak Mon Kuov, more drastic action was needed, the country's Prime Minister Chan Lin Yong said in a televised news conference. "We have decided that instead of tightening incrementally over the next few weeks, we should make decisive, targeted moves now to preempt escalating numbers of infections in these clusters," she said. "This is a frightening disease, but we must not panic in the face of it. Fear can sometimes make us do things that only make things worse – such as hoarding food, water and protective equipment like facemasks, or turning on certain groups that we want to blame for the outbreak. But Champosans are better than that. We know how to fight this virus. We must do it together."

The targeted lockdowns meant that more than 10,000 people living in the area surrounding Kompang Chom were effectively quarantined

– banned from leaving their homes for any reason other than to seek medical care or to buy essential provisions. From stockpiles maintained near Champosa's main airport, packages each containing 50 N95 masks were sent to every household in the city and its suburbs. Residents were instructed to wipe down the packages and leave them outside in sunshine and fresh air for at least three hours before opening them up. The same thing happened in Mon Kuov, where a population of just over 7,000 people was instructed to stay at home and avoid all but the most essential contacts with others. At the Mon Kuov clinic where the new confirmed and suspected cases were being monitored and where the 51-year-old had died, all staff and patients were put into a strict quarantine. No-one would be allowed in or out until at least two weeks had passed with no new confirmed cases, the hospital's director ordered.

The WHO's statement praised Champosa's swift response. Such urgent measures were justified, it said, given developments over the previous few days in the spread of the novel paramyxovirus. "These new developments are a cause for serious concern," the statement said. "While we know relatively little about the novel paramyxovirus, it appears to be a highly pathogenic virus in people, and is very likely to be a zoonotic agent. There is some emerging evidence that it may have a range of severity, with some patients experiencing relatively mild symptoms and others more serious ones. There is very limited but nevertheless important evidence to suggest that the virus may spread from person to person when an infected person has only very mild and initial symptoms – so-called pre-symptomatic transmission. But there is no evidence at this stage to suggest it can pass from person to person before an infected individual develops any symptoms at all – so-called asymptomatic or 'silent' transmission. The longer-term trajectory of the outbreak is unclear at this point, but the early trends are extremely unsettling. It is possible that the novel paramyxovirus could spread very quickly and cause many deaths. But it is also possible these outbreaks could be contained swiftly with non-pharmaceutical public health interventions such as isolation, quarantine, mask use and physical distancing. WHO scientists are working around the clock with experts in Champosa and the wider South Asia region to gain as much information and insight into the novel virus' origin, tropism and pathogenesis."

25TH JUNE 2027 (DAY 9)

Two days after its Prime Minister had put Kompang Chom and Mon Kuov into quarantine, and after all-night talks between government health officials in Champosa and officials at the WHO, at the G20 Presidency in Indonesia and at the United Nations, Champosa announced it was going into a full-scale nationwide lockdown. Several new cases of infection with the novel paramyxovirus had by then been detected at a clinic serving two other villages on the banks of the Pechong River, and Champosa's Prime Minister was afraid the disease could already be being carried across the country's borders. From 25th June, she announced, all passenger and cargo flights in and out of the country's airports would be grounded, and all major ports and railway hubs would be closed to passengers and freight. The only exceptions would be for incoming essential goods and services given Pandemic Prevention Status under new rules drawn up by the WTO as part of the Global Pandemic Treaty agreed by the world's governments in the wake of the Covid-19 pandemic.

"In close coordination with global health security organisations, the G20 Presidency, and United Nations agencies, we have decided to activate the emergency brake to stop all cross-border air, rail, ship and road travel from our country from today," Prime Minister Chan Lin Yong said in a televised address. Recognising that this would be a substantial and sudden hit for Champosa's economy, she also said she would be invoking a clause in the Global Pandemic Treaty that provided for immediate financial assistance to be offered to any country forced by the emergence of a Disease X threat to shut its borders and halt all but essential international trade. Champosa's decision was followed by scores of countries around the world moving to introduce restrictions on flights from so-far unaffected countries in Southern Asia. The United States banned all incoming flights from every country in the region, while Australia, Japan China, Argentina, Brazil and all members of European Union said they would not allow flights from six South Asian countries as a precaution against importation of the new virus. Several countries also imposed restrictions on international travellers coming in by sea, train or road. Visitors arriving into China, for example, would be screened and tested, and if positive for the novel paramyxovirus would have to remain in a government-approved hotel quarantine facility for 14 days.

26TH JUNE 2027 (DAY 10)

With cases of the new Pechong virus being reported across Champosa, G20 leaders convened on 26th June 2027 to vote on a request to the World Bank to open up a financing facility known as the Pandemic Prevention Facility, or PPF, to help national and regional authorities get ahead of any potential cross-border epidemic. The PPF was a substantial credit facility of $25 billion that had been agreed and secured in 2024 as part of the Global Pandemic Treaty. Designed as a highly-risk-tolerant fund for national governments, frontline responder organisations like UNICEF, for vaccine makers and for international agencies, it was able put big money up fast in the face of the newly-emerging viral threat.

Ahead of the vote, Indonesia's President addressed the meeting with a call to action and a reminder of how the pandemic of Covid-19 in 2020-2023 had exposed fatal gaps in the world's defences against emerging Disease X threats. "One of the most important things the Covid-19 pandemic reminded us of was the importance of diversifying vaccine production sites so that we can make new vaccines at speed and scale in any and all regions of the world," he said. "To prevent a repeat of Covid-19, and to avert the devastating impact it had on all aspects of society, from health and education, to prosperity, business and international trade, G20 countries via the World Bank have invested almost $7 billion a year over the past five years to boost vaccine production and health systems capacity in low and middle-income countries across the world. We have also reserved capacity at major pharmaceutical manufacturers in Asia, Africa and Europe – capacity that we can and should now activate to begin working with the developers of potential candidate vaccines against the novel paramyxovirus. This substantial preparation work and investment means we now have the ability to stop this new Disease X before it spills out into a pandemic. But we must act fast."

The G20 leaders meeting also featured a short intervention from Dr Richard Hatchett, Chief Executive of the Coalition for Epidemic Preparedness Innovations, who was asked to give his assessment of the Pechong threat. "Over several decades of worrying about, dealing with and preparing to prevent pandemics of emerging pathogens – collectively known as Disease X – I have rarely been as alarmed as I am today. It is no overstatement to say that the emergence of a novel paramyxovirus like the one identified in Champosa this month poses an existential threat to our way of life. We must do everything we can to ensure it does not become a pandemic," Hatchett told the meeting. "We have an extremely slim window of opportunity in which, if we are bold enough to take risks, we can neutralise this viral threat. None of us wants to see a repeat of the Covid-19 pandemic, with the dreadful toll it took on lives and livelihoods. And all of us know that if we had acted with more urgency and bravery back in January 2020, we could have saved millions of those lives and prevented some the catastrophic economic collapse and global destabilisation that followed. As well as speed, funding and leadership, we also know that vaccines are crucial to countering infectious disease threats. They are our most potent tool against pandemic threats and will be critical to our response to this new paramyxovirus. The faster we can develop and deploy an effective vaccine, the faster this outbreak – which is, in my view, an incipient pandemic – can be contained and controlled. Be in no doubt: this new virus is the next Disease X. Like the SARS-CoV-2 virus that emerged in 2019, it could easily run out of control in a most deadly way if we delay our response. To prevent global catastrophe, we must take the fight to it. And we must do that now."

The two speakers' words made an impact. Alarmed by the deadly and pandemic potential of the new Disease X threat, G20 leaders voted to open up financing of up to $20 billion to be set aside for at-risk development, manufacturing, and eventual procurement of six Pechong virus vaccines. In a joint statement issued that evening, they acknowledged the risks and the size of the outlay, but noted that speed was paramount in the face of the growing outbreak of the novel virus spreading from Champosa and into the wider Southern Asia region. “If the world acts together and with urgency to get the critical scientific research, development and technology transfer started against this new viral threat, we will reap the rewards in terms of lives saved, economic losses averted and, potentially, a pandemic prevented,” the statement said. “Stopping this deadly virus from spawning a wider epidemic or sparking a pandemic can only be achieved with swift and sizeable at-risk investments.”

Over the following two weeks, initial stage-gated grants of up to $10 million each from the Pandemic Prevention Facility went to the six candidate vaccine development teams, including the Giza Biologicals team in Hanoi and to a biotech company called Artemix Pharmaceuticals, based in Waterfall City, South Africa. Another initial tranche of PPF funding also went to the vaccine manufacturers that had agreed in 2024 to reserve up to 15 per cent of their production capacity to make rapid response vaccines for countries that needed them in the event of an epidemic of a new Disease X. The PPF funds allowed the manufacturers to put in advance purchase orders for raw materials, technology and training to be able to produce the new vaccines at scale and speed, even before the world knew if they would work, or would be needed. A third tranche of funding, of an initial $2 billion for immediate down-payments and up to $10 billion for follow-up payments as necessary, went to the Global Epidemic Vaccines Sharing facility, GEVAX, so that it could make advance purchase deals with manufacturers for any of the candidate Pechong virus vaccines that proved effective in trials and secured regulatory approval for emergency use.

17TH JULY 2027 (DAY 31)

On 17th July 2027, the scientist Aadhan Chakara woke up feeling excited, hopeful and slightly sick with nerves. This was the day he and his team at Vietnam’s Giza Biologicals were going to put the first dose of their newly-designed experimental Pechong virus vaccine into the arm of their first volunteer in their first human trial.

Because there were already reams of reliable data showing the vaccine platform’s high safety profile, this trial did not need to limit itself to just a handful of healthy young volunteers as an old-style Phase I trial would have done in the early 2000s. The new paramyxovirus vaccine candidate was essentially a modified version of a rapid response measles-vectored Nipah vaccine that had been developed and approved for outbreak use in 2026. Using that as a prototype, with a wealth of safety and dosing data behind it, meant Chakara's team could move their new vaccine straight to trials to test its efficacy – essentially to test whether it was able to prevent people from becoming infected with the novel paramyxovirus, and/or whether it prevented people who did get infected from becoming severely ill and dying. As such, just under 1,000 volunteers were screened and recruited to take part in the trial. They were people of all ages and from all the ethnic groups represented in the region's populations, from 18-year-old students to over-70s retirees. They had been called up from the region's reserve list – a pre-enrolled ready-to-go cohort of people which every country in the world had created in the wake of the Covid-19 pandemic. The reservists had volunteered to be called upon to be part of Pandemic Speed clinical trials in the event of a threat from a new Disease X.

In the past few weeks, Chakara and his Giza Biologicals team had designed, made and then tested the new vaccine in animal experiments using first ferrets and then a type of monkey called rhesus macaques. These earlier trials in macaques had been just as expected: the vaccine – created using an already well-tested and regulator-authorised platform technology – had not produced any serious adverse reactions in the monkeys. And immunogenicity results showed that after being given a single dose of the experimental new vaccine, all eight of the monkeys used in the trial developed protective antibodies against the Pechong paramyxovirus within 14 days. Chakara's team had used those small-scale studies also to zoom in on what they expected would be an optimum dose level for the experimental vaccine in humans.

At this stage, a month after the World Health Organization first alerted the world to the emergence of the novel paramyxovirus, the Pechong outbreak was posing a significant regional threat. One or several new vaccines that could protect people against infection or severe disease were looking like the only way a full-blown pandemic could be averted. And they were needed quickly. Having watched Champosan authorities move so rapidly and pre-emptively into a full-scale lockdown reminiscent of 2020, some neighbouring governments in countries where suspected cases were being picked up at border surveillance points had also opted to start introducing social restrictions. Like Champosa, they had learned from watching and assessing what places like South Korea and New Zealand had done in the early stages of Covid-19 pandemic to try and keep out and stem the spread of that novel virus. Hopeful that deploying similarly pre-emptive strategies would pay even greater dividends if they moved even more quickly this time, governments in Thailand, Vietnam and Malaysia had decreed that all schools from nurseries to senior education colleges should be closed as a precautionary and temporary measure from 30th June 2027, and that contacts of any confirmed and suspected Pechong cases should be traced, quarantined and tested for at least 14 days before being able to resume normal activities. Across these and several other populations in the region, gatherings of more than 12 people had been banned, work-from-home directives had been issued, large sporting fixtures were postponed and restaurants, bars, gyms, cinemas and theatres had been ordered to require mask wearing and proof of a negative test on entry, as well as to limit the number of customers to 30 per cent of their normal capacity to ensure people could stay at least two metres apart.

In Champosa, people had been living for three weeks under lockdown conditions that some critics had branded "draconian" but that had, for the most part, been accepted as necessary by members of the public. Nevertheless, financial pain, restlessness and impatience for a vaccine as a way out was already beginning to grow. Daily newspapers and radio shows featured calls from prominent business leaders and the head of the country's National Education Programme for the government to put a timescale on when vaccines might become available, and to set out its plans for vaccinating priority groups first. In a video address carried on national television channels and social media platforms, Champosa's Prime Minister Chan Lin Yong acknowledged and sympathised with the extremely difficult conditions her citizens were living with, but urged them to continue to make small sacrifices now to protect themselves, their families and the wider world from having to make what might be far greater sacrifices later. "This is a new Disease X, and we know from recent history how dangerous these novel pathogens can be to populations who ignore or belittle their potential threat," she told them. Working at home, wearing masks, and switching schooling to remote learning for a few weeks seemed a small price to pay to avoid that kind of horror again, she added. "If we make smart moves now – quickly and calmly – we can slow this outbreak's spread. That way, we will buy ourselves and other populations time to develop, test and deliver new vaccines with the ability to contain and hopefully extinguish this deadly epidemic."

Despite Champosa's efforts, and those of its neighbouring countries, the novel paramyxovirus's spread was not halted entirely. Cases of Pechong infection and death were rising across the country, and by mid-July, despite the stepped-up border controls, cases were also being reported in three neighbouring countries. On 17th July, the WHO's daily situation update on the epidemic reported 8,539 confirmed cases of infection with the novel paramyxovirus in four countries. The death toll stood at 412.

As Chakara's first half a dozen clinical trial volunteers arrived at the Hanoi University Clinical Research Unit, they too had been hearing and reading these latest reports. Some said afterwards they had felt apprehensive but at the same time proud to be stepping up to help what they saw as a critical effort to prevent Pechong from becoming a pandemic. As he rolled up his sleeve and proffered his left shoulder to the nurse administering jabs, the first volunteer, a 47-year-old bank branch worker from the Dong Da district of Hanoi, said he didn't mind if the injection was a bit painful. "It's worth it," he said. "If this new vaccine turns out to work, it could help save thousands, maybe millions of lives."

Chakara, who had not been able to come to the clinic due to restrictions on numbers, instead issued a media statement expressing his hopes for the trial: "Finding a safe and effective vaccine to prevent infection with the novel paramyxovirus is an urgent public health priority for this region and for the world," he said. "That's why this trial was launched at record speed as a vital step towards achieving that goal. It will involve hundreds of volunteers at this and three of our sister clinical research units in Jakarta, Kathmandu, and Kompang Chom – the epicentre of the outbreak. We are grateful to each and every one of them for joining us in this race to prevent a pandemic."

That same week in mid-July 2027, trials of two other potential Pechong vaccines, one developed by Artemix Pharmaceuticals in South Africa and another developed by a firm called Babylon Bio in the United States began at a series of centres in the Champosa region. For these clinical trials too, researchers focused on enrolling volunteers from cities, towns and villages where the disease had already, or might soon start to, spread. They were also keen to sign up volunteers who were also healthcare workers – and hence at higher risk of coming into contact with the virus. The closer the trials were to areas where transmission of Pechong was already high, the quicker they would be able to see whether or not their vaccines were providing protection.

26TH JULY 2027 – (DAY 40)

By the end of July 2027, the candidate vaccine makers – some of them major Big Pharma companies based in Europe and the United States and others smaller biotechs like Giza and Babylon and Artemix – had also begun sharing the "recipes" for their vaccines with large-scale vaccine manufacturing companies in Indonesia, South Korea, South Africa and Thailand. While it was too early yet to unblind any of the clinical trials to see whether those becoming infected with Pechong were in the placebo group or in the group given the new vaccine, the companies, backed by global health agencies, had decided to start ramping up production of doses just in case. This at-risk manufacturing was funded with grants from the PPF Pandemic Prevention Facility, which also committed the producers to selling the first 10 million doses each of the eventual vaccines, if they proved successful and were authorised for use, to the GEVAX vaccine sharing facility. The vaccines' price, as agreed in the 2024 Global Pandemic Treaty, would be little more than the cost of manufacture – coming to just under $5 per dose for the Giza Biologicals shot and around $8 per dose for the more mRNA-based vaccines. Maria Feemster, a Dutch health economist and former central banker who became the director of the Pandemic Prevention Facility when it was set up under the auspices of the World Bank in 2024, told a press conference that the pre-emptive manufacturing scale-up was a high risk but potential high reward decision. "By starting vaccine manufacturing scale-up immediately, the companies collaborating to create, develop and produce these potential Pechong vaccines can ensure enough doses are available as soon as possible," she said. "If the current clinical trials prove that any one, or all, of these new rapid response vaccines are safe and effective, as we hope, having doses ready immediately will be especially important for frontline healthcare workers, the elderly and vulnerable, and those with underlying health conditions."

At the same time as this move was made to kickstart mass at-risk production of potential Pechong shots, pharmaceutical regulators in Japan, The Netherlands and the United States made a joint announcement on a plans for an exceptional accelerated assessment and approvals process for vaccines being developed in an outbreak emergency. Rapid rolling reviews to evaluate the benefits of the new vaccines against the risk of infection or death from the novel paramyxovirus would be conducted in parallel with the clinical trials being conducted by Giza, Artemix and Babylon, the regulators said. Normally, data on a newly-developed vaccine's effectiveness, safety and quality would be collected from multiple stages of testing – from pre-clinical trials using animals right through to three or more stages of large-scale clinical trials in all age-groups and multiple populations – and then presented to regulators in one vast dossier. But the rapidly spreading and highly deadly Pechong outbreak demanded a different approach, the regulators said. "While there is still uncertainty about the transmissibility and virulence of this novel paramyxovirus, with more than 34,000 cases of infection now reported what is clear is that it poses a real and present threat to the people of Champosa and surrounding countries, and hence to all of us. Our decision to begin joint rolling reviews of these experimental Pechong vaccines will require intensive, accelerated and collaborative effort, but should ultimately benefit society and patients during this health emergency. It is based on preliminary results from laboratory and animal studies, as well as real-world data from the safe and effective use in people of Nipah vaccines developed on the same rapid response platforms. Those studies suggest these new vaccines could have the potential to reduce the ability of the novel paramyxovirus to replicate and multiply in the body, thereby preventing hospitalisation or death in patients with Pechong infection. We will swiftly and continually evaluate more data on the quality, safety and effectiveness of the vaccines as those data become available and until we have enough evidence to make a decision on authorising one or more vaccines for emergency use."

17TH AUGUST 2027 (DAY 62)

By the middle of August, despite the best efforts of Champosa and its neighbours to shut down borders, test, trace and quarantine all contacts of suspected and confirmed cases, and impose strict stay-at-home orders, the outbreak of the Pechong novel paramyxovirus had nevertheless become a regional epidemic. While the epicentre of the outbreak remained in Champosa, as did the vast majority of cases, other outbreaks of the disease had spread to six countries in South Asia, including two small clusters of infection reported in Singapore and Malaysia, neither of which had land borders with Champosa. A team of infectious disease outbreak experts from the WHO's Geneva headquarters had visited the region to investigate the epidemiological characteristics of the Pechong outbreak – including its source of infection, transmission routes, and the types and ages of people that were most susceptible. The WHO Pechong Mission was also there to find out more about the clinical manifestations of the new disease, especially the range of severity – how sick people were getting when they became infected. What they had found was alarming: with over 273,270 confirmed cases reported in six countries, so far, the death toll of 13,111 put Pechong's case-fatality rate at just under five per cent. Genetic analysis of thousands of samples from patients infected in the initial outbreak in Champosa suggested the disease appeared to have come from a species of bat common in the region called Pteropus fruit bats, and also known as "flying foxes." The pathogen might have been passed to humans via contact with contaminated bat faeces, the WHO experts said, or could in this case have made the jump into people via an intermediary host, possibly pigs. Epidemiological data suggested that no specific age-group was particularly susceptible to infection, but that the most severe cases of illness with Pechong and the majority of the deaths recorded by then had been in adults over the age of 50. School closures, work-from-home orders, contact tracing and quarantining measures appeared to be slowing the spread of the epidemic in all age groups, but not halting it completely. Detailed case study analyses of around 100 patients in the original starter cluster of infections at Kompang Chom's Sunshine Hospital pointed to an incubation period of three to 10 days, with an average of around a week. In most cases, the virus caused respiratory infection, muscle pain, fatigue and dizziness. More severely sick patients developed encephalitis – inflammation of the brain – that was in the most severe cases leading to seizures, coma and death.

The only good news – if you could call it that – in the WHO Mission's preliminary findings was that, as of then, no cases of Pechong virus infection had been reported in people who had been exposed to a Pechong positive patient who was not displaying any symptoms. This meant, the WHO said, that the novel paramyxovirus was unlikely to be spreading silently – or asymptomatically – at least that this stage.

26TH AUGUST 2027 (DAY 71)

There was good news and not so good news, too, from the clinical trials of experimental Pechong vaccines. Giza Biologicals' trial had gone swiftly from the start and had managed to enrol and dose close to 500 volunteers in the first two weeks with either the active or a placebo vaccine. With several of the Giza vaccine trial sites located the area of the outbreak's epicentre, the number of infections among trial participants was also ticking up. The trial's independent data board was monitoring the numbers every few days to assess when the right level might be reached to allow researchers to unblind data on the first few hundred participants. When that go-ahead was given, researchers would be able to see whether cases of infection with the novel paramyxovirus were more common in the placebo group than in the vaccine group.

Babylon Bio's trial, of an mRNA-based rapid response Pechong virus vaccine, was also proceeding at pace. One of its trial sites was near a hospital in the city of Mon Kuov in the west of Champosa where healthcare workers had been dealing with a steadily rising number of cases of Pechong infection, exposing them to the virus on a daily basis. Scores of doctors, nurses and other hospital staff who had not at first been on the list of pre-enrolled pandemic prevention trial volunteers had contacted the Babylon trial's chief investigators to ask to be allowed to take part.

For Artemix, though, the news was mixed. Its clinical trials had gotten off to a slower start, with some of the pre-enrolled ready-to-go cohort of volunteers at sites in Malaysia apparently not so ready to go after all. They were slower than volunteers in other places to respond to the call-up, and some didn't respond at all. The trial's recruitment team said some previously-signed-up volunteers had said they were reluctant to take part in this trial because the novel Pechong outbreak seemed, to them, to be relatively remote and unlikely to affect them. Then, at another of Artemix's trial sites – this time in Singapore where volunteer enrolment had been going a little more swiftly – an incident where a woman fainted and knocked her head on a desk just after she'd received the jab had spawned a spate of negative media reports and fuelled suspicions about the trial and the Babylon vaccine, further hampering progress.

In the meantime, public and political pressure was growing for regulators to fast-track a pandemic-prevention emergency use authorisation for any of the Pechong virus vaccines that could show they induced a protective immune response. Healthcare workers and some teachers in Champosa's major cities took to social media to call on the vaccine developers and the regulators conducting the joint rolling review to move as quickly as they could. "We are putting our health and our lives at risk to help care for patients with this pernicious disease," Darany Mae, a nurse working in the city of Kompang Chom said in a post on Instagram. "If any of these vaccines could offer us even some level of extra protection beyond the masks and PPE we're relying on right now, we'd want to get it as soon as possible." In response to these and other calls, the pharmaceutical regulators in Tokyo, Amsterdam and Washington who had agreed to collaborate on rolling reviews of the new vaccines issued a joint statement saying they were working "at pandemic speed" to complete risk-benefit assessments as the data from vaccine trials flowed in. "We already have plenty of evidence that the platform technologies used in the Artemix, Babylon and Giza candidate Pechong vaccines are extremely safe," the statement said. "As soon as we have enough data from tests on trial participants to be able to establish whether a vaccine is sufficiently immunogenic, we will move to decide on whether an emergency licence can be granted. We expect such an authorisation decision to be possible within a few weeks."

15TH SEPTEMBER 2027 (DAY 91)

Six thirty-one am on 15th September 2027 was the moment that Lita Kim, the paediatric consultant at Sunrise Hospital in Kompang Chom, briefly switched from being a provider of the latest available medical care to a receiver of the latest available medical defences. A nurse in the hospital's outpatient department sat her down in a small three-sided booth, asked her to roll up her sleeve, and administered the world's first approved Pechong virus vaccine, made by Giza Biologicals. The experience moved both doctor and nurse to tears. For Dr Kim, the weeping was partly due to exhaustion from week upon week working at the epicentre of the infectious disease outbreak that had landed on her hands when the feverish and fitful Boupha was brought into her hospital back in June. Nurse Darany Mae's tears flowed with relief for himself, his family and his country, that an end to the Pechong epidemic could be in sight.

"I feel so privileged to be the first person in Champosa, and the world, to be vaccinated with a Pechong virus vaccine," Dr Kim said. "It's the best reward I could get for the past few difficult and demanding months of fighting to give the best care possible to infected patients and help control the outbreak. We need to get this vaccine rolled out as quickly as possible – starting now with nurse Darany, who I will vaccinate later today, and then to all other healthcare workers and people at risk in Champosa and its neighbours. That way, we will stop this deadly epidemic in its tracks."

After a joint regulatory decision to issue both the Giza and Babylon vaccines with emergency restricted authorisations based on already established safety profiles coupled with new trial data showing both shots provoked an immune response to the novel paramyxovirus, Champosa's national emergency Pechong vaccination campaign took off swiftly. By the end of the first week – 22nd September – just under 80,000 people – many of them doctors, nurses, teachers and other public sector workers – had received a dose of either the Giza or the Babylon vaccine. Within a month, as continual monitoring of the initially vaccinated people showed no safety concerns and suggested they had heightened protective immunity, the number of Champosans vaccinated against the novel paramyxovirus had risen to more than 600,000.

The Pechong vaccination rollout was also beginning to gather pace in Vietnam, Thailand, Singapore and other affected countries. Initially, this rollout was with the mRNA shot made by Babylon and, a little later, with Artemix's viral-vector vaccine which had also been granted an emergency use authorisation. Being based in Boston, Massachusetts, Babylon had faced demands from the White House to fulfil a pre-emptive order for 500 million doses for the United States, despite no cases of Pechong at that stage having been detected in the Americas. But with mRNA production capacity reserved at several manufacturing sites in the Asia region under the terms of the funding from the Pandemic Prevention Facility, 600 million doses of the Babylon vaccine were already earmarked for the GEVAX globally fair allocation system which would deliver them to the most vulnerable populations in affected countries and others in the region most at risk in the epidemic.

With anxious and sometimes unruly queues forming at vaccination centres, Champosa's Prime Minister Chan Lin Yong made a video address seeking to reassure anyone still waiting to get immunised. "Fears about this deadly epidemic are completely understandable, but I must ask you to remain as calm as possible and to think about how to do the right thing for your families and for our country," she said. "I can reassure you that, thanks to the work of our National Vaccine Taskforce, 12 million doses of the Giza Pechong vaccine and just under a million doses of the Babylon shot have already been secured to cover our population. We hope to get the vast majority of those doses into arms within the next seven to 10 weeks and, in doing so, to dramatically bring down the number of new cases of this terrible infection."

TWO MONTHS LATER – 15TH NOVEMBER 2027

The World Health Organization's daily situational report on the Pechong virus outbreak on 15th November 2027 was very different from the more than 150 or so such updates it had issued since mid-June of that year. This report would be the last of the daily updates, it said, and the WHO team monitoring the outbreak would now switch to weekly updates that would track and document cases of the virus. The epidemic had by that stage in mid-November infected almost a million people in seven countries in the South Asia region. It had killed at least 31,830 of them. But, with the introduction and rapid rollout across 10 countries in the region of millions of doses of effective Pechong virus vaccines from Giza, Babylon and later Artemix, the region's epidemic curve had peaked and turned sharply south. Over the past two months, the numbers of cases had been waning fast. Death rates had come down even lower, with the vaccinations dramatically reducing the number of people who got severely ill with the disease. The Pechong paramyxovirus' potential for becoming a global pandemic had been thwarted. "It's important to recognise that the Pechong virus, while thankfully now largely contained, has not gone away completely," the WHO's emergency response director Cate Henderson said in a statement released with the final daily situational report. "We must remain vigilant, and we must alert others and act fast if there is even the slightest signal of a new flare up in the epidemic." The report said the WHO would continue to advise anyone in the region who had not yet been vaccinated, as well as travellers to Asia, to get themselves immunised as soon as possible. Enhanced surveillance in Champosa and across the region would also continue, it said, as would further strengthening of preparedness and readiness for Pechong or other infectious disease outbreaks across the world. "For the first time in history, and for the first time since the devastating Covid-19 pandemic of 2020 to 2023, humanity has used its ingenuity, quickly and collaboratively, to identify and get ahead of a new Disease X that was threatening to overwhelm the world and destroy millions of lives," Henderson said. "The novel paramyxovirus has put our collective pandemic prevention preparations to the ultimate test. The fact that we have not had a Pechong pandemic is testament to the power of prudent preparedness and quick-thinking leadership in thwarting infectious disease threats. To end pandemics forever, such prescience must now become the norm."

POSTSCRIPT

If it had taken 100 days to develop an effective vaccine against Covid-19, the first doses would have been available on 20th April 2020, when there were only 2.3 million confirmed cases of the disease. In fact, the first Covid vaccines were given on 8th December 2020, when more than 68.7 million cases of infection had already been confirmed in a vast and deadly global pandemic.[49]

In 2027, because we had made the investments in surveillance and scientific research, and because the world was alarmed enough and prepared enough to move at pandemic-busting speed and develop effective vaccines against the novel paramyxovirus within 100 days, the first doses were available when fewer than a million people had been infected. The epidemic did not spread beyond one region. There was no paramyxovirus pandemic. Potentially millions of deaths and trillions of dollars of economic losses were averted.

From now and into the future, it's truer than ever to say that while epidemics are inevitable, pandemics are a choice.

RESOURCES AND FURTHER READING

- Barry, J. M., *The Great Influenza: The Story of the Deadliest Pandemic in History* (London: Penguin Publishing Group, 2005)
- Crawford, D. H., *Viruses: The Invisible Enemy* (Oxford, UK: Oxford University Press, 2002)
- Farrar, J. and Ahuja, A., *Spike: The People vs The Virus. The Inside Story* (London: Profile Books, 2021)
- Feemster, K. A., *Vaccines: What Everyone Needs to Know* (Oxford, UK: Oxford University Press, 2018)
- Gates, B., *How to Prevent the Next Pandemic* (New York: Knopf Doubleday Publishing Group, 2022)
- Gilbert, S. and Green, C., *Vaxxers: The Inside Story of the Oxford AstraZeneca Vaccine and the Race Against the Virus* (London: Hodder and Stoughton, 2021)
- Giordano, P., *How Contagion Works: Science Awareness and Community in Times of Global Crises* (London: The Orion Publishing Group, 2020)

- Global Preparedness Monitoring Board report 2021, *From Worlds Apart to a World Prepared* (Geneva: World Health Organization, 2021)
- Gottlieb, S., *Uncontrolled Spread: Why Covid-19 Crushed Us and How We Can Defeat the Next Pandemic* (New York: HarperCollins, 2021)
- Harper, K., *Plagues Upon the Earth: Disease and the Course of Human History* (Germany: Princeton University Press, 2021)
- Hooper, R., *How to Spend a Trillion Dollars: The 10 Global Problems We Can Actually Fix* (London: Profile Books, 2021)
- Kenny, C., *The Plague Cycle: The Unending War Between Humanity and Infectious Disease* (New York: Scribner, 2021)
- Lewis, M., *The Premonition: A Pandemic Story* (London: Penguin Books Limited, 2021)
- Mackenzie, D., *Stopping the Next Pandemic: How Covid-19 Can Help Us Save Humanity* (London: The Bridge Street Press, 2021)
- Osterhaus, A. and De Pooter, D., *101 Questions and Answers on Influenza* (Maarssen, The Netherlands: Elsevier, 2009)
- Piot, P. with Marshall, R., *No Time to Lose: A Life in Pursuit of Deadly Viruses* (New York: W. W. Norton & Company, 2012)
- Schelling, T. C., *MicroMotives and MacroBehaviour* (New York: W. W. Norton & Company, 2006)
- Senthilingham, M., *Outbreaks and Epidemics: Battling Infection from Measles to Coronavirus* (London: Icon Books, 2020)

- Snowden, F. M., *Epidemics and Society: From the Black Death to the Present* (London: Yale University Press, 2019)
- Tregoning, J. S., *Infectious: Pathogens and How We Fight Them* (London: Oneworld Publications, 2021)
- Zakaria, F., *Ten Lessons for a Post-Pandemic World* (New York: W. W. Norton & Company, 2020)

ACKNOWLEDGEMENTS

This book would never have come into being without the effort, enthusiasm and encouragement of my colleagues at CEPI – most notably the Communications and Advocacy team led by the brilliant Rachel Grant, with Bjørg Dystvold Nilsson, Jodie Rogers, Mario Christodoulou, Rowena Madar and Tom Mooney – a great team of very kind and very clever people to whom I am extremely grateful.

Enormous thanks go, too, to CEPI's pandemic-worrier-in-chief, Richard Hatchett, who allowed me access to his emails, journals, personal messages and thoughts, and who gave up many early morning hours to share his experiences of the state of the world's preparedness for infectious disease outbreaks. Those hours of listening and talking were invaluable in shaping my thinking for *Disease X*.

CEPI's Melanie Saville, Dick Wilder, Joe Simmonds-Issler, Snorre Burmo, Frederik Kristensen, Nicole Lurie, Tim Endy, Raúl Gómez Román, Gabrielle Breugelmans, Mike Whelan, Sabrina Kriegner, Astrid Helgeland, Oyeronke Oyebanji, Matthew Downham, Barbara Ngouyombo, Luc Debruyne, Gwen Tobert, Jakob Cramer, Adam Hacker, Amy Shurtleff, In-kyu Yoon, Samia Saad, Abebe Genetu Bayih, Henshaw Mandi, Ingrid Kromann, Gerald Voss, Christopher Da Costa, Emma Wheatley and Nick Jackson all also gave me the benefit of their time and thoughts. Thank you.

The list of the many others at CEPI who supported and guided me is enormously long – stacked with people whose minds and hearts are full of brilliance and passion for making the world a safer and fairer place. Thank you all. I could not wish for a more inspiring group of people to work alongside. I am proud to be among you.

Thank you, also, to everyone outside of CEPI whose brains I picked – about book-writing or about pandemic-preventing, or about both. Some are quoted, while others are behind-the-scenes influencers without whom I would not have been able to make progress. I have benefited greatly from talking to you all: Seth Berkley, Peter Piot, Rajeev Venkayya, Erna Solberg, Ayoade Alakija, Jeremy Farrar, Osagie Ehanire, Chris Whitty, Fiona Fox, Anjana Ahuja, Sarah Gilbert, Charlie Weller, John Bell, Adrian Hill, Stanley Plotkin, Aaron Bernstein, Michael Wech, James Paton, Fergus Walsh, Francesca Barrie, Eddie Power, Mary O'Hagan, Brian Jordan, Tom Sheldon, Peter Tallack, Roger Highfield, Fiona Lethbridge, Charis Gresser, Maud Lardenois-Macocco, James Dray, Craig Breheny, Simeon Bennett and Etleva Kadilli, Carolyn Reynolds, Susan Hatchett and Luke Baker.

I am honoured that Sir Tony Blair wrote the foreword for *Disease X*. His strong belief that, if we properly and equitably prepare ourselves for outbreaks of infectious diseases, then we can fully and fairly protect ourselves against them, is one I wholeheartedly agree with. A big thank you also for the hard work and support of Adam Bradshaw at the Tony Blair Institute for Global Change.

Clare Grist Taylor, my agent and advisor at *The Accidental Agency*, has been a wise and wonderful presence throughout my writing of this, my first book. I am grateful for her belief in me and for her encouragement of me to find, and have confidence in, my own voice. Thank you, Clare. And thank you to Alister Doyle, my friend and former colleague at Reuters, who put me in touch with Clare.

So many more former colleagues at Reuters guided and supported me. I offer fond thanks to Julie Steenhuysen, Ben Hirschler and Stephanie Nebehay, all of whom I have thoroughly enjoyed collaborating with over the years. Heart-shaped thanks also to the great late George Short, who took me on as a rookie journalist in 1993 and gave me the start I needed to begin writing about the world.

My parents, Jim and Patricia, and my lovely sister, Annie, have been unwavering backers from afar as I have embarked on becoming a first-time author. Big love and big thanks to you all. And to those who have lived with me day to day while I set about planning and writing a book as well as training for an Ironman triathlon – Paul, Ben and Isabel – I am sincerely grateful for everything you are and do. Paul, thank you for your calmness and for listening and bearing with me when things got wobbly. Ben, thank you for being the best little boy I could wish for – always so lovely and positive, and never doubting I could do it. Isabel, thank you for being so kind and clever, and so very wonderful to have around – the best daughter in the world.

END NOTES

CHAPTER 1

1 Snowden, F. M. (2019). *Epidemics and Society: From the Black Death to the Present.* United Kingdom: Yale University Press.

2 https://covid19.who.int/

3 https://virological.org/t/novel-2019-coronavirus-genome/319

4 https://www.bbc.co.uk/news/in-pictures-51280586

5 Kundera, M. (2020). *The Unbearable Lightness of Being.* United Kingdom: Faber & Faber.

6 https://www.who.int/director-general/speeches/detail/who-director-general-s-opening-remarks-at-the-media-briefing-on-covid-19---11-march-2020

CHAPTER 2

7 https://news.harvard.edu/gazette/story/2015/11/an-indictment-of-ebola-response/

8 https://www.gov.uk/government/news/over-70s-and-at-risk-brits-advised-against-travelling-on-cruise-ships

9 https://twitter.com/BilldeBlasio/status/1234648718714036229

10 https://cepi.net/news_cepi/cepi-to-fund-three-programmes-to-develop-vaccines-against-the-novel-coronavirus-ncov-2019/

11 https://www.cgdev.org/article/new-study-covid-19-vaccine-rollout-fastest-global-history

12 https://www.medrxiv.org/content/10.1101/2020.03.03.20029843v3

CHAPTER 3

13 https://www.nber.org/system/files/working_papers/w26847/w26847.pdf

14 Crewe, D., Green, C., Gilbert, S. (2021). *Vaxxers: The Inside Story of the Oxford AstraZeneca Vaccine and the Race Against the Virus.* United Kingdom: Hodder & Stoughton.

15 https://www.ft.com/content/67e6a4ee-3d05-43bc-ba03-e239799fa6ab

16 https://committees.parliament.uk/publications/7496/documents/78687/default/

17 https://www.who.int/docs/default-source/coronaviruse/transcripts/who-transcript-emergencies-coronavirus-press-conference-full-13mar2020848c48d2065143bd8d07a1647c863d6b.pdf

18 https://www.exemplars.health/emerging-topics/epidemic-preparedness-and-response/covid-19/south-korea#author

19 https://www.who.int/docs/default-source/coronaviruse/who-china-joint-mission-on-covid-19---final-report-1100hr-28feb2020-11mar-update.pdf

20 Ahuja, A., Farrar, J. (2021). *Spike: The Virus Vs. The People – the Inside Story.* United Kingdom: Profile.

21 https://www.ox.ac.uk/news/2021-11-23-former-vaccine-taskforce-chair-calls-fundamental-reset-government-systems-and

22 https://www.theguardian.com/business/2020/oct/13/imf-covid-cost-world-economic-outlook

CHAPTER 4

23 Foege, W. H. (2011). *House on Fire: The Fight to Eradicate Smallpox.* United Kingdom: University of California Press.

24 https://apps.who.int/iris/bitstream/handle/10665/55594/WHF_1998_19(2)_p113-119.pdf

25 https://cepi.net/wp-content/uploads/2021/09/Proposal-to-establish-a-globally-fair-allocation-system_March-25_2020.pdf

26 https://gisaid.org/hcov19-variants/

27 Jacqueline M. Fabius, Nevan J. Krogan, Creating collaboration by breaking down scientific barriers, *Cell*, Volume 184, Issue 9, 2021, ISSN 0092-8674, https://doi.org/10.1016/j.cell.2021.02.022

CHAPTER 5

28 https://www.regjeringen.no/contentassets/d0b61f6e1d1b40d1bb-92ff9d9b60793d/no/pdfs/nou202220220005000dddpdfs.pdf

29 https://www.bbc.co.uk/news/uk-politics-56361599

30 https://www.bbc.co.uk/news/uk-54182368

31 https://www.thetimes.co.uk/article/48-hours-in-september-when-ministers-and-scientists-split-over-covid-lockdown-vg5xbpsfx

CHAPTER 6

32 https://www.niaid.nih.gov/news-events/experimental-hiv-vaccine-regimen-ineffective-preventing-hiv

33 https://quod.lib.umich.edu/c/cohenaids/5571095.0488.004?rgn=main;view=fulltext

34 https://www.who.int/publications/m/item/draft-landscape-of-covid-19-candidate-vaccines

35 Harris, J. E., The repeated setbacks of HIV vaccine development laid the groundwork for SARS-CoV-2 vaccines. *Health Policy Technol.* 2022 Jun;11(2):100619. doi: 10.1016/j.hlpt.2022.100619. Epub 2022 Mar 21. PMID: 35340773; PMCID: PMC8935961.

36 Gates, B. (2009). *Business @ the Speed of Thought: Succeeding in the Digital Economy.* United States: Grand Central Publishing.

37 https://www.un.org/development/desa/dpad/wp-content/uploads/sites/45/WESP2022_CH3_SA.pdf

CHAPTER 7

38 https://www.gpmb.org/annual-reports/overview/item/2019-a-world-at-risk

39 https://iccwbo.org/content/uploads/sites/3/2021/02/2021-icc-summary-for-policymakers.pdf

40 Krammer, F. Pandemic Vaccines: How Are We Going to Be Better Prepared Next Time? *Med (N Y).* 2020 Dec 18;1(1):28-32. doi: 10.1016/j.medj.2020.11.004. Epub 2020 Dec 5. PMID: 33521752; PMCID: PMC7836605.

41 https://www.reuters.com/world/europe/germany-decides-principle-buy-f-35-fighter-jet-government-source-2022-03-14/

42 Hooper, R. (2021). *How to Spend a Trillion Dollars: The 10 Global Problems We Can Actually Fix.* United Kingdom: Profile.

43 AlRuthia, Y., Somily, A. M., Alkhamali, A. S., Bahari, O. H., AlJuhani, R. J., Alsenaidy, M., Balkhi, B. Estimation of direct medical costs of Middle East Respiratory Syndrome Coronavirus infection: a single-center retrospective chart review study. *Infect Drug Resist.* 2019 Nov 7;12:3463–3473. doi:10.2147/IDR.S231087. PMID: 31819541; PMCID: PMC6844224.

44 https://www.thelancet.com/journals/langlo/article/PIIS2214-109X(18)30346-2/fulltext

45 Bernstein, A. S., Ando, A. W., Loch-Temzelides, T., Vale, M. M., Li, B. V., Li, H., Busch, J., Chapman, C. A., Kinnaird, M., Nowak, K., Castro, M. C., Zambrana-Torrelio, C., Ahumada, J. A., Xiao, L., Roehrdanz, P., Kaufman, L., Hannah, L., Daszak, P., Pimm, S. L., Dobson, A. P. The costs and benefits of primary prevention of zoonotic pandemics. *Sci Adv.* 2022 Feb 4;8(5):eabl4183. doi: 10.1126/sciadv.abl4183. Epub 2022 Feb 4. PMID: 35119921; PMCID: PMC8816336.

46 https://www.gpmb.org/annual-reports/overview/item/2020-a-world-in-disorder

CHAPTER 8

47 https://www.pnas.org/doi/full/10.1073/pnas.2105482118

48 https://www.gpmb.org/docs/librariesprovider17/default-document-library/gpmb-annual-report-2021.pdf?sfvrsn=44d10dfa_9

POSTSCRIPT

49 https://covid19.who.int/

Index